Before You Build

Before You Build

A Preconstruction Guide

by Robert Roskind

Introduction by Ken Kern

Ten Speed Press

1⊜
TEN SPEED PRESS
P O Box 7123
Berkeley, California 94707

Library of Congress Catalog Number: 81-51897
ISBN: 0-89815-036-1

Book Design by Hal Hershey
Cover Design by Brenton Beck
Illustrations on cover and throughout book by Jon Larson

4 5 — 92 91 90 89 88

To my parents, my sister, Susan, my brother-in-law, Alan, and my nephews, Marc and Adam, who opened their homes and hearts to me as I wrote this book.

And to the light that shines within us all.

Contents

Introduction

Who knows but if men (sic) constructed their dwellings with their own hands, and provided food for themselves and families simply and honestly enough, the poetic faculty would be universally developed, as birds universally sing when they are so engaged.

THOREAU

BUILDING ONE'S OWN home is a work experience that can provide not only shelter, but as Thoreau perceived, can augment the urge for creative expression. In recent years, the staff of the Owner-Builder Center has provided thousands of prospective home builders with instruction and information about dwelling construction. House building becomes an easier and more creative task for those intent on direct participation in the making or remodeling of their own shelter.

As the cost of land continues to inflate and as interest rates and energy costs rise, more and more individuals and families are unable to afford contractor-built housing. The only alternative for these people is to build or remodel for themselves, even though owner-building often appears puzzling and formidable.

House construction represents the largest single expenditure of time, money, and resources that most people will make in a lifetime. Without sound preparation and careful attention to detail, novice home builders may not only squander their investment, but also suffer nightmarish agonies of frustration and self-doubt.

This need not be the case. Building one's own home can be executed with efficiency and a sense of real accomplishment when actual construction is preceded by adequate preparation. To this end, owner-builders have long expressed their need for a comprehensive guidebook and house building checklist; even professional builders require a similar breakdown of their operations. Checklists, requiring detailed analysis of a proposed project, help one avoid the often perilous adventure that house building can sometimes be.

With the Owner-Builder Center Preconstruction Guidebook, owner-builders may aquire a well-founded approach to house design and planning. A skillfully designed directory, this builder's guide can disclose costly items of which many of us are either ignorant or tend to view as inconsequential. In short, this guidebook and checklist is an educated attempt to help self-builders avoid the pitfalls of that phase of house building that precedes the on-set of construction work. From the gleam in one's eye to stark reality, the owner-builder will find that little of importance is overlooked in this text, which is as perceptive as it is thorough.

KEN KERN

I WOULD LIKE to acknowledge the following people for their help in providing information for this book:

Owner-Builder Center Students:

Glen and Sandy Brown
Jose Dorado
Morris Ho
David Kook
Alan Leavitt
Harry Reich
John Tompkins

Other Builders:

Gene Walter and Sidney Robinson
Cathedralite Domes of Reno
Reno, Nevada

Ron Mickens
High Country Domes
Montrose, Colorado

Marc Durand
Domes of Sonoma
Santa Rosa, California

Jan Vander Linden
South Eastern Cathedralite Dome
North Augusta, South Carolina

Thomas Wood
Condor Domes
Ventura, California

Jerry Brizendine
JB Dome Homes
Wichita, Kansas

Donn Marinovich
Mother Lode Domes
Twain Harte, California

Charlene Thompson
Colorado EcoDomes, Inc.
Golden, Colorado

James M. Smith
Smith and Associates
Chapel Hill, North Carolina

Consultants:

Passive Solar:

Jim Maloney
Jeff Poetsch
Dave Roberts
Don Elmer

Owner-Builder Center Staff:

Bill Youngblood
Mike Hurley
Dave Roberts

ROBERT ROSKIND

The Workbook and How to Use It

THIS BOOK IS an accumulation of the experience of myself, a professional builder, many of the Owner-Builder Center staff, other contractors around the country, and many of our students. Given this input, I feel confident that if all the questions in this book are answered you will have few, if any, surprises awaiting you. When the questions weren't answered, we have seen surprises cost thousands of dollars, hundreds of hours work, and at times render the project unworkable. This book has been written so that you will be able to avoid all this. Its intention is to inform you what questions you need to ask, help you arrive at the answers, and organize the data in an orderly way.

Completing this work book is an essential and orderly way to begin your project. This should help you determine the feasibility of the project, as well as alerting you to any hidden problems. It is suggested that the entire book be studied and work pages be completed before: 1) purchasing the property, 2) siting and designing the house, and 3) making the final decision to go ahead with the project.

Consider your answers carefully, as your decisions will be based on them and correct decisions at this stage may save you time, energy and money; perhaps even spare you future disappointments and/or problems. It is suggested that you try to answer all the questions, although this may not always be possible. Many questions are apparent by their asking, others are footnoted with information at the end of each chapter which may clarify the question and aid you in answering it. We suggest that you read the footnote even if you understand the question, as the footnote may contain material that is new to you. For questions that need still more explanation, a chapter by chapter bibliography has been included. Understand, also, that you may encounter special situations or areas not covered in this workbook. For this reason ample space has been provided for you to add your own questions and notes.

It is important that you be able to identify the major issues—those which could effect the feasibility of the entire undertaking—and have them answered to your satisfaction prior to beginning work on your project. To aid in this we have highlighted the essential questions in boldface type, though we still advise that you answer them all.

Do not be intimidated by the amount of information you need to process even before beginning construction. It is only the beginning but you are capable of doing it and doing it well. If, after reading this book, you decide not to build, it has completed its purpose since it is best to realize in the beginning that this is something you were not prepared to do. Should you decide to continue and either build or contract your own home, this book will help you get off to a good start. Building or contracting your own home is a tremendous commitment, probably greater than you realized. It needs to be well thought out, planned, and timed. It will demand more from you and your family than you ever thought and should not be entered into without much careful consideration. Be willing to decide not to do this; it is a major life-decision.

Because building or contracting your own home demands so much, it also has the capability of giving even more. The experience offers not only financial benefits, but also a potential for self-growth rarely found in other endeavors. It will push you to your limits until they are no longer your limits and in the end you will have a house that may feel more like "home" than any other possibly could.

It is our intention in offering this workbook to make your decisions and journey easier, and to help you more quickly reach your goal—a loving home for you and your family.

Good Luck!

ROBERT ROSKIND
The Owner-Builder Center

1/Buying the Land or Lot

After you have evaluated the land you wish to own from an objective viewpoint, determined that you can afford it, and decided that it has most of the features that you are seeking, ask yourself, "Is this really the piece I want?" and then be totally honest with your answer.

THE IMPACT of the property that you choose on the success of the entire project and especially on your life in the house once it is completed cannot be emphasized enough. There are three main considerations at this stage. First, be sure that the property you buy is suitable and workable for the project that you have in mind. The questions and information in the following chapter will help you in ascertaining this. We have talked to many people who have spent thousands of dollars on a piece of property only to find out after the purchase that for one reason or another they could not build or do what they wanted on that piece. Sometimes no structure at all could be built!

With today's energy costs sometimes being greater than monthly mortgage payments, a second vital consideration for the potential land buyer concerns the solar and natural ventilation qualities of the land. Any house built today that does not use these design elements is obsolete before it is built.

Both the heating and cooling loads on a home can be reduced 50–80% by proper design and orientation and this begins with the proper choice of the land and then the site. I remember a conversation I had with Barbara Kern, Ken's wife, as to why they were going to move. Her response was "We never could do anything with this piece; it's on the northern slope."

In the following chapter, *Choosing the Site*, there is a section on Solar Heating, to assist you in making a study of your property and site as far as microclimate is concerned. It is best to complete this study before purchasing the land, as your site orientation, design, and microclimate will have an ever increasing impact on your monthly fuel bill.

The third consideration is that once you find the property that has all the features you need, you are certain that it is really the piece you want and feel good about. If the land does not feel welcoming and appealing to you, it will not enhance the quality of your home life. Too often owners have chosen property by its price, solar orientation, location, or some other *physical* feature of the property, or out of weariness of the search for the right piece, only to later realize that they never really feel quite at peace with the land and home. An extra six months of looking or an additional thousand dollars will fade into the background over the many years you will live in the house. Trust that when you find the right place you will know it and your decision will bring you clarity and joy.

If you are buying land on which to build a speculation house for resale, the way you feel about the land is not so vital. As a real estate friend of mine once said: "Only three things matter when selling a home: location, location, and location."

After locating a piece that you are seriously considering buying, it is usually wise to work out a deal with the seller, sign the papers and pay the deposit, but have the final purchase contingent on certain requirements being met. These can be as definite as locating water, getting a building permit or variance, or as broad as your approval of the preliminary title report, which would allow you to cancel the sale for any reason, without explanation, after reviewing the preliminary title report. These contingency clauses tell the seller that you are serious enough to sign papers and give a deposit, yet allow you time to check to be sure the property will work for you. These clauses usually run from thirty to ninety days. If the seller does not want to grant one there may be something about the property he or she does not want you to know.

For Further Reading

Finding and Buying Your Place in the Country. Les Scher. New York: Collier Books, 1974.

The Owner Built Homestead. Barbara and Ken Kern. New York: Charles Scribner's Sons, 1977.

Buying the Land or Lot

Location of land____________________

Parcel # ____________________

Owner____________________

Address ____________________

Phone ____________________

Asking Price $ ____________________

Terms ____________________

Size ____________________

Real Estate Agent____________________

Phone ____________________

Does the land welcome you? ☐ YES ☐ NO

Is this really the land that you want? ☐ YES ☐ NO
If not, why are you buying it?

____

____

____

Can you afford the cost? ☐ YES ☐ NO
If not, what are your options?

____

____

____

____

Is there a good southern exposure for passive or active solar use?[1] ☐ YES ☐ NO

Can this exposure be later blocked? ☐ YES ☐ NO

Are solar rights available and enforced? ☐ YES ☐ NO

Is it enough land for your purposes? ☐ YES ☐ NO

Is there enough land to meet minimal size requirements? ☐ YES ☐ NO

What are the required distances between well and septic tank? ______________

Between well and house? ______________

Between septic tank and house? ______________

Between house and road? ______________

Are all these possible on your land? ☐ YES ☐ NO

What is the zoning on the land? (Check with local zoning office.)[2]

____

____

____

____

Are there any special provisions in the local codes that could prohibit you from building what you want? □ YES □ NO

(If yes, what are they, and are variances possible?)[3]

____

____

____

Are there any homeowner's associations or architectural committees that have placed restrictions on the land?[4] □ YES □ NO

If yes, what are the restrictions?

____

____

____

What is the zoning of the neighboring land?

____

____

Are any of the following planned for your area:

Power plants □ YES □ NO

Large roads □ YES □ NO

Recreation areas □ YES □ NO

Missile sites □ YES □ NO

Subdivisions □ YES □ NO

Other ______________________________

Are any code or zoning variances needed in order to build on the land? □ YES □ NO

Who will be responsible for acquiring these? ______________________________

Are there any special restrictions due to fire regulations? □ YES □ NO

____

____

Are there any existing permits accompanying the land? □ YES □ NO

Can the property be subdivided? ☐ YES ☐ NO

If so, how?[5]

____

____

____

Is there a possibility the subdivision law will be changed? ☐ YES ☐ NO

Will this effect your plans for the land? ☐ YES ☐ NO

Who owns the neighboring lands and what are their plans?

	Owners	*Plans*
N	____________	______________________
S	____________	______________________
E	____________	______________________
W	____________	______________________

What are the projected growth rate and zoning changes in the area?

Growth Rate:

- ☐ Light
- ☐ Medium
- ☐ Heavy

Zoning changes: ______________________

Are any power plants or large roads planned? (Consult local planning office) ☐ YES ☐ NO

Person contacted ______________________

What are the land taxes? $______________

Home taxes? $______________

Are these likely to increase? ☐ YES ☐ NO

Are there any local bond issues? ☐ YES ☐ NO

What are the setback restrictions:

Side yard ______________________

Front yard ______________________

Back yard ______________________

Special conditions ______________________

Are there any building moratoriums in the area?[6] ☐ YES ☐ NO

Is the title free and without liens and complications?[7] ☐ YES ☐ NO

Name of title company ______________________

Person contacted ______________________

Results of preliminary title report: ______________________

____

____

Are there any existing easements on the property? ☐ YES ☐ NO

If so, what are they?[8]

____

____

____

Will you need any easements for roads, utilities, etc. to use your land? ☐ YES ☐ NO

____

____

____

What is the cost of the title search? $______________________

Title insurance? $______________________

What are the closing costs of the sale? $______________________

Are there any mechanics' liens on the property?[9] ☐ YES ☐ NO

Is the paperwork (options, terms, offer, escrow, etc.) clear and legal? ☐ YES ☐ NO

Will any attorney be needed? ☐ YES ☐ NO

If so, who will pay?

☐ Buyer
☐ Seller

Will the owner subordinate if you are not paying cash?[10] ☐ YES ☐ NO

Are there other lenders on the land? ☐ YES ☐ NO

Will they subordinate to construction lenders? ☐ YES ☐ NO

Will the terms of the land purchase interfere with the mortgage loan?[11]
□ YES □ NO

Is the survey marker easy to locate? □ YES □ NO

Will a survey be necessary?[12] □ YES □ NO
If so, what will the cost be? $________________
Who will pay, buyer or seller?

□ Buyer
□ Seller

What percentage of the purchase price is going to the realtor? ____________%[13]

Can you find banks that will loan on unimproved lands? □ YES □ NO

Can a larger parcel be deeded a few acres at a time in order to get bank financing?[14]
□ YES □ NO

Could you buy a larger parcel and get a cheaper price per acre? □ YES □ NO

What are the values of neighboring land and houses?

____

____

Are there any existing improvements? □ YES □ NO

□ Well
□ Roads
□ Septic
□ Power
□ Phone
□ Buildings
□ Pond sites
□ Building area

What are the conditions of the roads and bridges to your land?

□ Poor
□ OK
□ Good
□ Very Good

Is the road large enough and in good enough condition for large trucks (e.g., concrete trucks)?[15]
□ YES □ NO

Who will maintain the roads and bridges to the land?

Individual owners? ______________________

Owners' cooperative? ______________________

Local government? ______________________

Other? ______________________

Are there any special uses of the road (e.g., transport of hazardous materials)?

____

____

Will the road be too rough on your car? ☐ YES ☐ NO

What are the visibility conditions from the road to the house during the different seasons?

Spring ______________________

Summer ______________________

Autumn ______________________

Winter ______________________

Is snow removal on the roads available? ☐ YES ☐ NO

Is local road very busy or traffic very fast?

Winter ☐ YES ☐ NO

Summer ☐ YES ☐ NO

Who is legally liable for the roads and bridges?

____

____

How much will electricity cost to be brought to the property?

Underground $______________________

Overhead $______________________

Person contacted at power company ______________________

Phone number ______________________

Is natural gas, propane, or fuel oil needed and available? ☐ YES ☐ NO

Is there a history of power outages? ☐ YES ☐ NO

How much will it cost to bring in a phone line? $______________________

Are any of the following alternative energy systems feasible?

Wind □ YES □ NO

Pelton Water Wheels □ YES □ NO

Solar □ YES □ NO

What are the soil conditions at your building site?[16]

Type of soil ______________________________

Load bearing capacity ______________ lb./sq. ft.

If weak soil, will an engineered foundation be needed? □ YES □ NO

Name of soil engineer ______________________________

Phone ______________________________

Fee ______________________________

Name of structural engineer (to design foundation, if needed) ______________

Phone ______________________________

Fee $______________________________

What is the soil like for growing?

____

____

Does the soil perc?[17] □ YES □ NO

Person doing perc tests ______________________________

Phone ______________

Cost $ ______________

Test results ______________________________

Can you do your own perc test? □ YES □ NO

Will areas other than your choice sites also perc? □ YES □ NO

What would be the cost of an engineered septic system, if one is needed?

$______________

In subdivisions, do promised water and sewage hookups really exist? □ YES □ NO

If not who will pay for them? ____________________

____

Do any of the following conditions exist?

□ Land fill
□ Potential slides
□ Coastal plains
□ Diseased trees
□ Insect problems
□ Noise problems
□ Pollution problems
□ Flood plain

Comments: ____________________

Will foundation be expensive due to:

□ Slope
□ Land fill
□ Bedrock
□ Boulders
□ Expansive soil
□ Poor Drainage
□ Soil Slippage
□ Earthquakes
□ Flood Plains
□ Frost Line
□ Accessibility (of trucks and equipment)

If a soil report is needed, can a previous one from a neighboring house be used? □ YES □ NO

Is good, potable water available at a reasonable price?[18] □ YES □ NO

Depth of #1 neighbor's well ____________

Depth of #2 neighbor's well ____________

Depth of #3 neighbor's well ____________

Estimated depth of your well ____________

Well digger's name ____________________

Phone ____________

Estimated price $____________

What is the response time for the following:

Fire ____________________

Police ____________________

Ambulance ____________________

Cab Service ____________________

How far to the:

Hospital ____________________

School ____________________

Store ____________________

Lumber Company ____________________

Gas Station ____________________

Swimming hole ____________________

Bus line ____________________

Dumpster ____________________

Place of Worship ____________________

Will distance from land to supplies result in added delivery charges? □ YES □ NO

Is garbage collection available? □ YES □ NO

What is the monthly cost? $____________________

How far to the closest fire station?[19] ____________________

Is fire insurance available? (Note: mortgage companies require this.) □ YES □ NO

How far is it to your place of employment?____________________

How much time does it take to get there? ____________________

How many children live in the area and what are their ages? (This can be very important to your children.) □ YES □ NO

Is there a place to build a pond or lake? □ YES □ NO

Will government agencies assist with your lake building? □ YES □ NO

Are there any natural building materials available?

- ☐ Trees
- ☐ Rock
- ☐ Sand
- ☐ Adobe clay
- ☐ Gravel

Other ______________________________

Is there a possibility of purchasing neighboring property? ☐ YES ☐ NO

____

____

____

Has the land been sprayed with pesticides? ☐ YES ☐ NO

Is logging going on in the area? ☐ YES ☐ NO

Is the land on a flight pattern from a local airport? ☐ YES ☐ NO

Is there much hunting in the area? ☐ YES ☐ NO

What, if any, are the restrictions on tree cutting?

____

____

____

How much unusable or unwanted land must be bought with your property?

____

____

If you are planning to sell the house, is the land located in an area where houses sell easily and in the price range you are planning to ask? ☐ YES ☐ NO

What is the local weather like?

Number of days of sun ____________

Number of days of rain ____________

Number of days of snow ____________

Amount of fog ____________

Direction of Prevailing winds:

Summer ____________________

Winter ____________________

Average Temperature:

June ____________________

Sept ____________________

December ____________________

March ____________________

Do you feel at home in the community? ☐ YES ☐ NO

Will your design match the land? ☐ YES ☐ NO

Have you spent enough time on the land to be sure you want to buy it? ☐ YES ☐ NO

To chose the site? ☐ YES ☐ NO

Are duplexes or triplexes allowed? ☐ YES ☐ NO

Are rental units allowed? ☐ YES ☐ NO

Are play areas for children available? ☐ YES ☐ NO

What kind of wildlife is on the property?

____

____

____

What kind of floral life is on the property?

____

____

____

Are there any yearly dues in the subdivision? ☐ YES ☐ NO

What is included with this yearly fee?

____

____

____

Are the building sites 25% grade or more? (If so, building is discouraged.)[20] □ YES □ NO

Are there any special insect problems on the land? □ YES □ NO

____

____

Will drainage or erosion problems occur due to construction (cuts into banks, parking or road areas)? □ YES □ NO

____

____

Is the land located downstream from a dam that could offer potential danger? □ YES □ NO

Is newspaper delivery available? □ YES □ NO

Is spring water delivery available? □ YES □ NO

After absorbing all this data, what are your feelings about buying this land?[21]

____

____

____

____

____

____

____

____

____

____

____

____

____

____

Some points on this checklist may take some time to answer (water, perc test, etc.) and you may want to tie the land up with an offer that is contingent on the answers to some of these questions.

Notes

1. Any house built today which doesn't use the sun as a heating source will be obsolete in many parts of the country. Passive solar design is proven, feasible, and economically available in all parts of the United States. Unlike active solar heating, which uses lots of equipment and money, passive solar heating uses good house design and site orientation to allow the house itself to both store and collect the heat. It is simple and it works and can provide 30–85% of your heating needs. All of this is just beginning to become widely known.

If you are planning to build a solar home, the land must be evaluated as to its ability to provide enough sunlight to make solar heating possible. In the site chapter I have provided a checklist for site selection in regards to solar. You may want to fill this out before purchasing the land. Also, because the land is not optimal for solar does not always mean it is useless for a solar home. It may simply be a problem site that will require a little more imagination and innovation to adapt to solar. A good solar designer may be needed to assist you in evaluating your site's solar value.

2. Many urban, rural, and suburban areas have zoning ordinances and zoning departments in order to regulate the type of development in any particular area. This is to assist in providing a good mix or separation of commercial, industrial, and residential development. Ascertain before buying the property not only if you can build and do what you want on the property, but also what can be built on the neighboring pieces and in the general area. Often areas without zoning can be problematic. I knew of an owner-builder who built his house only to have a chicken farm with 300,000 chickens move onto the land directly across the road. Needless to say, the stench, noise, and visual impact did little to enhance the quality of life in the area.

Though zoning has the force of law it also has limitations. Often zoning ordinances can be contested. One of the first court contests on zoning was in the US Supreme Court case of *Euclid vs Ambler Realty Company* in 1926. The court cautioned that:

> "The constitutional right of the owner of property to make legitimate use of his lands may not be curtailed by unreasonable restrictions under the guise of police power. Thus the owner will not be required to sacrifice his rights absent substantial need for restrictions in the interest of public health, morals, safety and welfare. Where zoning restriction exceeds the bounds of necessity for public welfare, it must be stricken as an unconstitutional invasion of property rights."

If zoning ordinances are restricting your plans in any way, question whether they are reasonable and *enforceable*.

3. Many code and zoning ordinances can prevent you from building certain kinds or even sizes of buildings. Check with the local authorities for any restrictions. Often the height of a building will be limited or minimum size requirement may exist. Also, if you plan to do any sort of business: cottage industry, or manufacturing in your home, this may have to be approved. Agricultural pursuits or animal raising may be prohibited. Even activities as simple as home day care, or selling products to friends through your home may be restricted.

4. In subdivisions or developments, certain restrictions regarding use of the property, style, or size of the buildings are often attached to the property. These restrictions, called "CC and R" (covenants, conditions and restrictions), have the power of law and owners who do not follow them can be sued by the developer or development association. These can include such items as: care of the grounds and buildings, size and style of buildings, setbacks from property lines, type of roofing and siding that can be used, color of exterior paint, use of the land, number of pets, volume of exposed noise, insecticides that can be sprayed on the property, size of trees that can be cut, firearms that can be used, use of common grounds, number of structures that can be built on the property, how the power lines can be run, etc. They can be very simple and easy to comply with, or they can be as restrictive as to dictate a lifestyle. Be sure you feel comfortable with any that you may be asked to live with.

5. You may have plans to subdivide your property in hopes of selling some of it to help pay for the entire parcel, or to speculate, or to sell to friends or family that you would like to live near. If any of these are your intention, be sure it is allowed and that the cost is not prohibitive. These actions can become very political and can often become very much of a community issue. Sometimes roads, sewer systems, power, and street lights may have to be installed and elaborate maps and permits obtained in order to divide land. Even purchasing large parcels to split into two equal pieces on which to build two separate residences may require much money and energy to subdivide. Very often the division of land at all, no matter what the size, is prohibited.

Warning: If you plan to divide your land at any time in the future and your local ordinances will allow you to do this you may be wise to have the land divided as soon as possible. If they later change the subdivision laws and make them more prohibitive it may be more difficult, if not impossible, for you to realize your plan.

6. In order to retard growth, for any reason, municipalities will sometimes enact building moratoriums. Sometimes these may not directly restrict building, but apply to something that, in essence, halts building—moratoriums on septic tanks, sewer hookups, water hookups, etc. This can last for years, make your property worthless, and tie up capital that you could use elsewhere to buy a home. Moratoriums can be enacted at any time by a referendum from the community or by policy enacted by the governing bodies. Be sure one does not exist in the area you wish to build in and that there is little possibility of one being enacted soon. If there is a possibility, you may want to quickly apply for your building permit, since building moratoriums are not usually retroactive if you have your permit. I know of a family that bought a parcel of land worth $30,000.00 in Half Moon Bay, California. Soon after the purchase, a building moratorium was enacted and now after four years they are still unable to build and potential buyers are few.

7. If the title or deed on the property is not free and clear, your ownership of the land may be in question even after the house has been built. You can avoid this by checking the records at the courthouse for any possible problems with the deed. You may have the title searched by a professional title company. Usually this costs several hundred dollars, but is an essential expenditure since you can risk losing your home and land. After the title search, which will be required by most lenders, you are given a guarantee that the title is clear of all liens and encumbrances, mineral and timber rights, etc. Also the title company can issue an insurance policy for the price of the property and the house, which will cover you should the title company have missed anything and your ownership come in question. This is a one-time expense of another several hundred dollars and again it is advised. Usually the cost of both of the services is paid by the buyer.

8. A previous owner of the property may have granted easements on the land to someone. These would then be attached to the deed and all future owners would be required to recognize them. Someone could, for instance, show up at your door twenty years after you had built your house and lived on the land and inform you that they are ready to harvest all the trees on the property, the rights to which were sold to them twenty-two years before by the former owner—you never noticed in the small print in the deed (this happened to a friend of mine). The fact that those trees are now part of your back yard may have little impact.

Easements may be for timber or mineral rights or for access over your property, either by a utility company or a landowner who needs to cross your land to get to the road. Often, a strip of land adjoining any road is attached by easement to utilities or to the county to maintain the road. A good title search should reveal any easements on the property. Read the entire deed thoroughly and be sure that you understand it all before you finalize the deal.

9. Many states have mechanic's lien laws. Basically, this allows any workers or materials suppliers who have furnished either labor and/or materials to a job site to place a lien on the property if they have not been paid. If any work has been done on the piece you are considering buying there may be a lien on the property or one may be placed on the land after you have purchased the property (liens can be placed from 30–120 days after the work was done). The kind of jobs that may cause liens include: tree clearing, road or bridge work, grading, septic tank or well installations, temporary power pole installations, pond building, etc. If any of this work or any work at all has been done recently on the property, you may want to get lien release on the property from the suppliers and workmen. Also, many lending institutions will require this before qualifying the project for a loan.

10. Subordination by the land owner is crucial if you plan to finance the construction of your home. It works like this:

Say you buy a piece of land for $20,000.00, paying $10,000.00 down and owing the seller of the land the rest to be paid in monthly installments over the next five years. The seller then keeps a deed of trust on the property for the remaining $10,000.00. If you fail to pay the monthly payments, the seller can have the land sold, take what he or she is owed from the selling price, and return what remains to you after all expenses are paid. Even if you have built a $100,000.00 house on the land you owe only $10,000.00 on, it can be sold if you default on the note. You can begin to see why banks and lenders would not lend money to build the $100,000.00 house if it could be sold to pay someone else. In essence, they become second in line for payment should you default, and this is seldom acceptable to them.

If you plan to owe on the land, and hope to borrow money to finance the house, obtain a subordination clause from the seller before you purchase the land. Basically, the clause says that the seller will "subordinate" his interest in the property to another lender, usually one who is providing financing for the house. This puts the bank or lending institution first in line should you default. Many sellers are very willing to subordinate and it happens often. Usually, they know that the land and house will be worth enough to pay both themselves and the lending institutions.

If you can not get a subordination clause, you may have to pay off the note on the entire land before you can find financing for the house. Sometimes the lending institution will lend you the money for this; at other times they will require that you own the land free and clear before they loan on it.

11. Sometimes the agreements in the land purchase in some way interferes with the process of receiving construction and mortgage monies. Before closing your land deal, talk with your lender about whether the deal could in any way interfere with your future loans.

12. Often, the survey markers are hard to find or may not be there at all. In this case, a survey is a good idea. The cost of the survey is often paid by the seller, or split between the seller and buyer. Surveys can be quite costly. There have been times that I have walked property with a real estate agent and later discovered we were not looking at the proper piece. Because of the cost of a survey, property owners are often reluctant to have one done until a sell is certain and often they may not even be certain of the property they own. We recently designed a house to fit on a small lot only to find that the property lines were not where we originally thought, and many changes in the design were then needed.

13. If the land can be bought directly from the owner, instead of going through a realtor, the initial down payment frequently will be considerably less, since the entire realtor's commission is attached to the down payment. This way of buying land has its advantages of a lower down payment, dealing directly with the seller, and a lower overall cost. Its disadvantages are that you will not have the expertise that the realtor brings to the transaction in your behalf. Also, the paper work for a real estate deal can get rather complex, even for a professional, and this is dealt with by the realtor as part of his commission. So there are reasons for doing it both ways.

14. You may be buying a large piece of property and want to build on the property before you have paid on the entire piece. Perhaps the seller of the land does not want to give you a subordination clause and the bank will not finance your building project unless you own the land clear or have a subordination clause. What you can do is deed part of the land off from the rest (if this is allowed through your local zoning) and that section alone is released free and clear by the owner of the land so that a loan for the home is possible. Usually some lump sum of money passes hands at this point. Parcels can continue to be released as you need them, if you are developing the property.

15. The conditions of the roads can become very important, both during and after construction. If the roads are too steep, or bridges too weak, the concrete and lumber trucks may not be able to get to the building site, or they may have to carry half loads. Both of these situations can cost you much time and money. After the house is completed, a rough road can damage a car very quickly, causing another expense. Before you buy the land, check out road conditions.

16. The type of soil in your area may have some bearing on the way your foundation must be built. If there are soil problems on your site, your foundation may have to be specially constructed and may cost several thousand dollars. If the soil is weak, like loose clay or land fill, or is expansive, like adobe, or is loose sand, it will not be able to hold much weight and the foundation will have to be custom designed and built. Usually your local building office or contractors can tell you if there are soil problems in your area and, if so, what is usually done about them. Find out about foundation problems before purchasing land; they can be costly.

17. If central sewer systems are not available, individual septic tanks and fields are provided for each house. All the waste water from all the bathroom and kitchen fixtures are drained into a large underground tank. The solids remain in the tank where they are dissolved by enzymes. The liquid waste drains through underground perforated pipes into the soil which acts as a filtering system (the leach field). If the soil is too compacted, too wet, or consists of too much clay or rock, it will not allow the system to work properly and will cause a potential health hazard.

The ability of the soil to drain these waste waters is its ability to 'perc'. A perc test is performed in the area where the septic tank and leach fields will be to see if the soil will perc well enough for a septic system to work properly. Basically, a hole is dug in the soil down to the level of the bottom of the leach field. The hole is then filled with water. The drop in the water level in the hole is then measured over a certain period of time. If the water filters through the soil well enough, the water level in the hole will drop a required amount in the allotted period of time. If the soil has problems, the water level will drop very slowly. Usually your local health department will give you the details for the test: how wide the hole must be, how deep, and how many inches the water must fall in how many hours. They will often require that one of their inspectors check the water level. Some areas may not have this requirement, as the soil in that area offers no problems to septic tanks.

If your soil will not perc properly and a septic tank is required, you may need to build an engineered septic system. This can cost many thousands of dollars and necessitate trucking in dirt that will perc. Another option, which is more ecologically sound and also less expensive, is to use a waterless toilet, such as the Clivus Multrum. These systems, which have been used successfully in Sweden for many years, are only recently being introduced to the United States. They compost waste products in a large tank that is stored in the basement, underground, or on the first floor of a two story building. The system is odorless, clean, and uses a toilet that looks much like those we now use. They use no water or energy and supply bacteria-free compost every few years. As of now they are legal only in a few areas, such as Maine and areas of California (in California, you must also provide a standard flush toilet although you do not need to use it). However, the product is proven safe and effective and permission to use the system in your area may hinge on someone presenting the research evidence to your health official and arguing the case. Since other officials have approved use of the Clivus Multrum, there is already a precedent.

The Clean Water Act of 1977 will assist builders and owners in providing adequate and environmentally sound sewage disposal systems. The legislation encourages design, construction and operation of innovative and alternative sanitation systems. You can use this act not only to assist you in getting approval for an alternative system but perhaps for receiving funds to implement the system as well.

Some highlights of the Clean Water Act:

1. Requires all grant applicants to show why the recycled water alternative has not been selected.
2. Authorizes funding for alternative treatment that is not the most cost-effective as long as the life-cycle cost doesn't exceed the most cost-effective substitute by more than 15%.
3. Provides 100% funding for research projects that demonstrate "innovative technology."
4. Provides federal grants that would pay 85% of the alternative or innovative systems' construction costs.
5. Grants are available for privately owned treatment systems serving one or more existing residences or small commercial structures.

18. Water is usually supplied to homes by one of a number of different methods: municipal water systems; cooperative water systems; individual deep wells; individual spring systems; individual shallow wells; surface water from creeks, streams, or lakes; cisterns for rain water. Whichever system you are planning to use be sure you know the regulations regarding use, and the initial cost of hooking-up to or installing the system, before you buy the land. This is especially true of wells, which can cost from $2–10,000.00. For more on this, see the chapter on water systems.

19. The distance to your local fire station may be a determinate factor in receiving fire insurance on your home, which is a must for obtaining a mortgage. You should also think of protecting your investment. With the growing use of wood as a main heat source, and wood interiors becoming more popular, the chance of a house fire is also increasing. If it is too great a distance to the local fire station it may endanger your ability to get a loan, as well as your investment. If the distance and response time is too great you may want to install a sprinkler system or other fire deterrent system.

20. The following is a quote from "Residential Site Planning Guide" by the National Association of Home Builders: "Slopes of 0 to 4 percent are visually flat. Land that has a 4 to 8 percent slope is developable with flatland units. Slopes of 8 to 15 percent are developable with many types of units. In many places minor streets can slope up to 15%; however in colder regions with severe ice and snow problems, builders should check with local codes for maximum slopes. If land slopes from 15–30% it is developable with specially designed hillside units and possible use of an inclinator or hillside elevator. Streets with slopes greater than 15% are usually considered hazardous to traffic. Land that slopes more than 30% is generally not developable except with special design concepts."

21. Much in this chapter has been devoted to obtaining and digesting data in order to assist in making your decision about purchasing a piece of property. You may want to consider that the most important piece of data is you and your family's feeling about the property, especially after digesting all the objective data that may have a bearing on the feasibility of your project. Does the land or lot welcome you? Does the idea of buying the property excite you, or are there still serious reservations? Is it where you want to live for a long period of time? When creating a successful home these feelings are as relevant as the price and solar orientation of the land and they are often overlooked.

Additional Information

Additional Information

2/Choosing the Site

Remember that your home is not only the house but the land surrounding it as well.

AS IN purchasing the land, siting the house has three main requirements for success: that it possess most of the physical features and qualities that will make the project workable; that it be suitable for solar orientation and natural ventilation so that these can be incorporated into your house design; and that you feel like this is the spot you want to live and have direct everyday access to—that it feels right to you. If any one of these is missing it can affect the success of the project in many ways.

The selection of the site is often determined by the property, its setbacks and size, or the fact that there are only one or two possible building sites. Other times, there may be many possible sites to choose from. In either case it is advisable to complete this chapter *before* buying the land, to be sure there is a suitable site on the property. The study of proper siting is quickly becoming a science, especially with the rising cost of fuel bills.

This chapter has been divided into two sections to help you evaluate the site. The first section, *Choosing the Site*, is concerned with construction problems at the site and what life in the house will be like. The second section, *Solar Heating and Cooling*, will help you evaluate the site regarding the energy efficiency of the building. Spend some time on your potential sites, at different times of the day and year if possible, and try to imagine living right there. If one spot is right for you, you will feel drawn to it.

Some pieces of land should never be built on, yet sellers and real estate people will seldom admit that. At times, I look at San Francisco, with its steep hills and shifting soil, on top of one of the most active earthquake faults in the world, and I think "There should never have been a city built there." Some sites are like that too, maybe not because of an earthquake fault, but perhaps due to drainage, view, trees, soil, or the like. Sometimes you can just look at a site and know how well it would accommodate a home; other times more investigation is needed. Problems are not always apparent and can be discovered when it is too late. This section will assist you in evaluating your potential site in terms of construction of the house and life on the site.

For Further Reading

The Food and Heat Producing Solar Greenhouse. Bill Yanda and Rick Fisher. Santa Fe, NM: John Muir Publications, 1980.

Passive Solar Architecture: Logic and Beauty. David Wright and Dennis A. Andrejko. New York: Van Nostrand Reinhold Co., 1982.

The Passive Solar Energy Book. Edward Mazria. Emmaus, PA: Rodale Press, 1979.

The Solar Home Book: Heating, Cooling and Designing with the Sun. Bruce Anderson and Michael Riordan. Brick House Publishing Co., 3 Main St., Andover, MA 01810, 1976.

Your Low Cost Energy Efficient Shelter for the Owner and Builder. Eugene Eccli, ed. Emmaus, PA: Rodale Press, 1976.

Choosing the Site

What is the direction of the prevailing wind?[1]

Summer ____________________

Winter ____________________

List the size, type, and location of all natural conditions near your site:

Hills ____________________

Trees ____________________

Shrubbery ____________________

Creeks ____________________

Rock Formations ____________________

What is included in your views?

N ____________________

S ____________________

E ____________________

W ____________________

What are the soil conditions at the site?

□ Poor □ Good

□ Okay □ Don't know

Is there good natural drainage?[2] ☐ YES ☐ NO

What is the wet season water table? ______

Could there be drainage problems in the future from houses built uphill from you? ☐ YES ☐ NO

Will your building create drainage problems for houses downhill from you? ☐ YES ☐ NO

Are there any geological hazards nearby? ☐ YES ☐ NO

If so, what are they? ______

Trees:

Tree No. 1

Height ______

Location ______

Fast growing ☐ YES ☐ NO

Diameter ______

Health ______

Species ______

Deciduous (do they lose leaves)? ☐ YES ☐ NO

Age ______

Tree No. 2

Height ______

Location ______

Fast growing ☐ YES ☐ NO

Diameter ______

Health ______

Species ______

Deciduous (do they lose leaves)? ☐ YES ☐ NO

Age ______

Tree No. 3

Height ______________________

Location ______________________

Fast growing ☐ YES ☐ NO

Diameter ______________________

Health ______________________

Species ______________________

Deciduous (do they lose leaves)? ☐ YES ☐ NO

Age ______________________

Tree No. 4

Height ______________________

Location ______________________

Fast growing ☐ YES ☐ NO

Diameter ______________________

Health ______________________

Species ______________________

Deciduous (do they lose leaves)? ☐ YES ☐ NO

Age ______________________

Are the trees to the south deciduous? Or evergreen?

____

____

____

Are there any dead or diseased trees that could endanger the house?[3] ☐ YES ☐ NO

Is there a location downhill from the house that would be good for the septic system?[4] ☐ YES ☐ NO

Will the neighbors' wells or septic tank locations influence the placement of yours and thereby affect the location of the house? ☐ YES ☐ NO

What are the required setbacks in your area?[5]

Front Yard ______________________

Back Yard ______________________

Side Yard ______________________

What is the slope of the land?[6] ______________%

Will any grading be necessary? ☐ YES ☐ NO

If so, how much? ______________________

Name of grading contractor ______________________

Phone ______________ Cubic yds. to be moved ______________

Location dirt will be dumped ______________________

Does the site where the land fill is going to be dumped need to be inspected or engineered? □ YES □ NO

What are the available energy sources:

Wood	□ YES	□ NO
Solar	□ YES	□ NO
Wind	□ YES	□ NO
Electricity	□ YES	□ NO
Propane	□ YES	□ NO
Natural Gas	□ YES	□ NO
Coal	□ YES	□ NO

How far will materials need to be carried to the site?[7] ____________________

Is a contour map of the site available through any of the local government agencies? □ YES □ NO

Will the road or any road easements influence your choice of home sites? □ YES □ NO

Will the existing or planned location of utility lines influence the choice of site? □ YES □ NO

Will a road be difficult to build to the site? □ YES □ NO

If yes, why? ____________________

____

Have your vehicular and pedestrian pathways been worked out? □ YES □ NO

Are parking areas well planned? □ YES □ NO

Will headlights of approaching cars shine into the house? □ YES □ NO

Will any large trees need to be removed on or near the foundation trenches that will create large holes to be filled?[8] □ YES □ NO

If so, how large? ____________________

How many? ____________

Can large machinery such as concrete trucks, lumber trucks, and cranes get access to the site? □ YES □ NO

Are there any possible changes that could influence your choice of site? □ YES □ NO

If so, what? ____________________

____

Is there enough privacy? □ YES □ NO

How far to the closest fire hydrant? ______________

How does the site feel to you?

____

____

____

If noise is a problem are noise breaks possible, such as trees or earth mounds?
□ YES □ NO

Are you making the best use of your properties? □ YES □ NO

Is your site-use plan well thought out? □ YES □ NO

What will your yards be like if you build on the site?

____

____

____

____

Notes

1. Much attention has been given in the past few years to solar heating, especially passive solar which uses the house and the site design to heat the building. Recently, attention has begun to turn towards passive solar cooling. Again this begins with good site and house design. Before elaborate house designs are implemented the site needs to be thoroughly evaluated to see how its features and orientation can be used to keep the home at a comfortable summertime temperature. Wind flow and shading are essential and often homes that are well shaded from the hot summer sun and have adequate air flow are comfortable without air conditioning or other passive solar cooling techniques.

If summer heat is a problem in your area be sure to note the direction of winds and location of shade trees (these can be planted) in choosing your site. It is said that if the State of California had planted shade trees on the west side of all the homes in Sacramento to cut off the hot summer sun, it would save more energy than is produced by the local Rancho Seco Nuclear Power Plant.

2. Remember the flattest piece of land may not be the best one to build on due to its lack of drainage. A better lot may be one that has slope so that it will drain properly. Steep sites can also have a serious drainage problem as the surface and ground water running down the slope may cause erosion. These sites may require drainage tiles around the foundation.

Also remember that whatever drainage patterns are on the site before the house is built will drastically change once a house has been put on the land. Some of our earlier students built a house in North Carolina on a very flat piece of land that had poor drainage. Within a year the floor system developed a wood fungus that threatened the entire house. The foundation had to be retrofitted with vents, the area below the house needed to have channels dug in the dirt to provide drainage for the excess water, and drainage will be an ongoing problem because the site was not well selected with regards to drainage. Some sites are meant never to have houses built on them.

3. Be careful of that beautiful old oak that convinced you to buy this piece of land so that you could build your house under it; it may one day fall through the roof. One of our summer projects in Nevada City, California, was a beautiful 1500 sq. ft. house that was designed to wrap around three trees that we wanted to keep both for their aesthetics and summer shading. The entire house design was based on the location of these three trees. One tree had to be cut and to our surprise the entire inside was rotten even though the tree looked healthy. We are now wondering about the health of the other trees. Also, the needed excavation for the foundation has exposed and cut some of the roots making the situation even more precarious. We know of other students who bought five acres of pine forest and built their home there. Now, due to the powder post beetle, their beautiful pine forest is a lawn of stumps. Trees are a wonderful asset to a site both for beauty and shade, but they are alive and therefore, like all of us, vulnerable to change.

4. If the septic tank or sewer system is uphill from the house, a special ejector toilet will be required to flush uphill and all other plumbing fixtures will have to be fitted with special equipment to push the waste water uphill to the tank. These are troublesome and expensive.

5. Be sure to find out how the setbacks relate to the roof overhangs and porches and decks. Sometimes these are allowed to encroach into the setback areas, other times they are not. Also find out whether you can obtain a variance from setback rules should your design and location make it difficult to comply. These are frequently more lenient rules than some of the others.

6. As a contractor, I would never estimate the cost of building a house without first visiting the lot. I would look closely at two things to evaluate their impact on the time it takes to build: accessibility of supply trucks to the house, and slope. Because of the high cost of land, more and more people can now afford only steeper lots that once may have been passed over. Not only can foundation costs soar to $10–25,000.00 on steep lots, but the construction of the entire house becomes much more time consuming and labor intensive. Scaffolding and ladders have to be erected, materials have to be lifted, people have to move slower due to the danger of falling. All of this slows the process down considerably. All houses eventually reach a phase of building where the work can no longer be done from the ground, but those on steep lots reach this phase sooner—the heights are higher and the scaffolding and ladders more difficult to erect. Also, accessibility around the site is reduced and movement made more difficult. Slope and access can have a major impact on the building process and on your time, labor, and money budgets. Don't overlook them.

7. The distance materials have to be carried may seem like a small matter, but it can tremendously influence the building process. If supply trucks cannot get close to the site, all of the materials will have to be carried in, which adds hours to each work week. Few people really understand the amount of time, energy, and persistence it takes to build a house unless they have already built one. If materials have to be carried in to the building site, it does not mean that the project is not feasible, only be sure that you understand that you are adding another element of time and labor to an already immense task.

8. If large trees are to be removed on or near your foundation line, be careful. Their root systems also need to be removed, leaving a large hole that must be filled for the foundation to be stable. This can be costly. If the tree is not removed, it can also cause problems if the root system starts to uproot the foundation. In either case, be sure you have thought out the presence or removal of trees near your foundation.

Additional Information

Additional Information

SOLAR HEATING AND COOLING

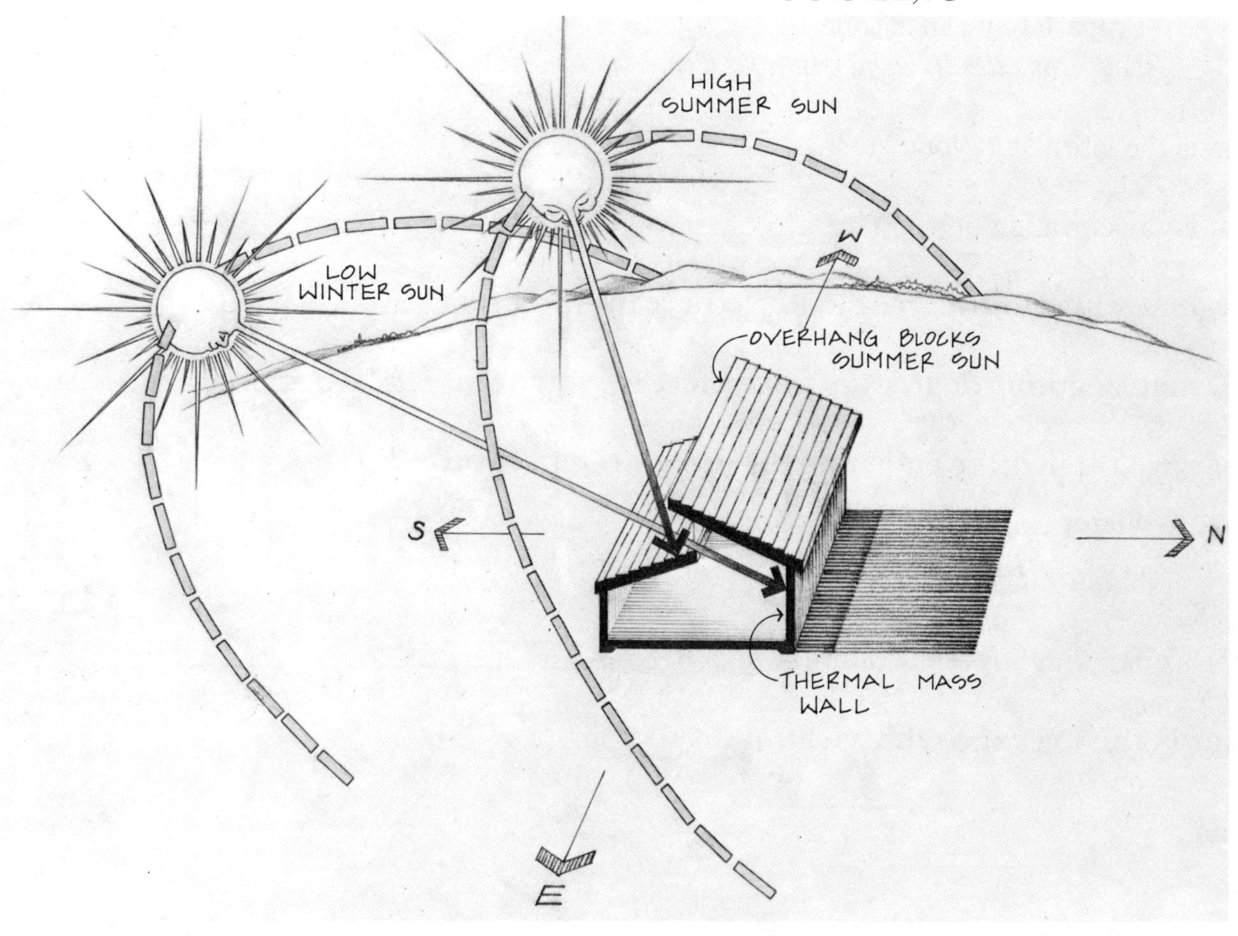

THIS SECTION is intended to assist you in evaluating your site regarding passive solar heating, active solar water heating, and passive solar cooling. In moderate and cool climates, using passive solar designs is a must for escaping from ever increasing energy bills. In moderate and hot climates, or cold climates with hot summers, passive solar cooling should be given consideration.

Passive solar heating, unlike active solar heating, requires no equipment, panels, or hardware, only proper design, glazing, mass, and orientation. It is simple and straightforward and requires minimal initial capital outlay. Though not all sites will be ideal, many which appear to have little solar value may just be problem sites and still workable for passive solar design, though placing more demands on the designer. Some sites are of no value for solar use at all; I would consider passing by these.

This part of the chapter is divided into two sections written by two solar professionals. The first section is by Dave Roberts, solar consultant for the Owner-Builder Center; and the second section, which deals with both passive solar cooling and heating, is by Don Elmer of Lawrence Berkeley Laboratory, Passive Cooling Program. Both sections are written in somewhat different formats and there is minimal overlap between authors.

Solar Heating and Cooling

Prepared by Dave Roberts
Solar consultant, Owner-Builder Center

What is the latitude of your site? ____________________

What is the elevation of your site? ____________________

How many heating degree days are there in your area?[1] ________________________

How many cooling degree days are there in your area?[2] ________________________

What are the winter and summer design temperatures?[3]

Winter ____________________

Summer ____________________

What are daily/nightly temperature swings?[4] ____

What is the average relative humidity?[5]

Summer ____________________

Winter ____________________

Is any insolation data available? □ YES □ NO

Will the winter water table affect solar heating mode of your house?[6] □ YES □ NO

Are your neighbors willing to cut or trim any obstructing trees?[7] □ YES □ NO

What location on your land receives the most mid-day sunlight?[8]

____

____

Is this location best for house or garden?[9]

House ____________________

Garden ____________________

Which is more important in your area? Heating or cooling?

□ Heating
□ Cooling
□ Both

If your site is on a hillside, which direction is the slope?[10] ______________________

Which direction is *true* south (not compass south)?[11] ______________________

Is a topographical map available? □ YES □ NO

Are there any obstructions? □ YES □ NO

What are the solar rights in your state?[12]

____

____

How fast growing are the trees near the site?[13] How old are they?

____

____

What type of trees are they?

____

____

From what direction do the winter storms come?[14] ______________________

Where is the dominant view for most glazing?[15] ______________________

What type of ground cover is on the south side?[16]

____

Does the ground cover change in summer and winter? □ YES □ NO

Will the branching pattern of a deciduous tree block much winter sun? □ YES □ NO

Where is the most shaded area of your land?[9]

____

Where are the street or common areas in relation to your site?[17] ______________________

____

Are there other solar homes in your area that you could visit? □ YES □ NO

Will seismic considerations in your area make a mass wall dangerous or prohibitively expensive?[15] □ YES □ NO

On which side of your home will your black top driveway or parking area be?[16] ___________

Will CC and R in any way dictate your solar design?[12] (See Note 4, page 16.) □ YES □ NO

Do you have access to a good wind area where you could put a windmill, i.e., the top of a hill or a valley? □ YES □ NO

Can the landscape be changed to funnel wind where you want it? □ YES □ NO

____

____

Are you willing and is it necessary to remove your trees for good solar design?[7] □ YES □ NO

____

How is the land to the south zoned?

____

____

____

Notes

1. The heating degree day is a measure of the severity of cold weather in any area. It measures how far below 65 degrees the average temperature is. For each day when the average temperature is below 65 degrees, degree days are recorded. When these are added up for a year, the resultant number is the number of heating degree days for the location. For example, if on a given winter day the temperature averaged 37 degrees, the number of degree days accumulated for that day would be (65–37) 28 degree days. It is useful as a relative measure of different climates. The San Francisco area has approximately 3000 degree days during the year, while Minneapolis is blessed with an 8400 degree day climate. Most parts of the country fall between these two different figures, and the number that is attached to your area will let you know what you must plan for in terms of insulation levels, plantings, etc.

2. A cooling degree day is computed like a heating degree day, but it measures how far above the standard temperature the average daily temperature is. San Francisco has no cooling degree days, thanks to the coastal fog. Some of California's interior valleys can have thousands of cooling degree days. A high number of cooling degree days should indicate a balanced approach to the cooling/heating system. A strong imbalance in the ratio of cooling degree days to heating degree days should promote one approach, either heating or cooling, over the other.

3. Winter and summer design temperatures are close-up views of the climate that degree days do not supply. They indicate a percentage of days when a certain temperature limit is exceeded. A typical listing would give a percent, typically 1%, 2.5%, or 5%. The listing would be followed by a minimum or maximum temperature, perhaps 35 degrees or 97 degrees. These would be the outer limits of the temperature range in your area, and the temperature for which the system should be sized. Most heat loss calculation techniques use this number to arrive at a maximum hourly heat loss figure.

4. Another close-up view of the climate you are in is the daily/nightly, or diurnal, temperature swing. The Sacramento area in California has uniformly high summer daytime temperatures, but usually has cooler temperatures and breezes at night. This diurnal temperature swing is a valuable design tool that can allow you to balance your house's temperature midway between the extremes, or to pull your house's average temperature towards the high or low extreme depending on your inclinations.

5. In addition to temperature, relative humidity is a key factor in determining human comfort. High humidity limits your ability to cool air in the summer, and low humidity leads to health problems in cold winter climates. Learn how to control your interior environment to create comfort.

6. Heat storage systems that are ground-based must be isolated from the surrounding water or the heat will be lost. If you are in an area which has a high water table you should plan alternatives to ground-based storage (slabs).

7. A solar system must have at least four hours of sunshine a day to be effective. If your prospective site is shaded by a neighbor's trees, it will not work for a solar home unless the trees are cut. Ask your neighbor if she/he is willing. Some ordinances are being enacted that declare trees that shade solar systems to be a public nuisance. Help get them enacted in your area. (P.S. Plant some shorter fruit trees in exchange.)

8. The requirement for four hours of sunlight in a day helps narrow the search for your house site unless you own a large pasture or meadow. If you live in a forest and love each tree like a brother don't bother doing a solar system. Just super-insulate.

9. If you have only one clear sunny spot you must decide whether it is best for the house or the garden. What are your priorities? Do you have many cooling degree days? Do you have wood to heat the house? Do you like to garden? Could you be satisfied with a south side greenhouse attached to the house?

10. The slope of the land determines the amount of sunshine that will fall on the ground and house. A south facing slope is going to be much warmer than a north facing slope. Keep in mind the relative heating/cooling needs of your climate when you are looking at different slopes. A steep north facing slope can block the winter sun completely.

11. Magnetic south is not solar south for most of the United States. The easiest way to tell true solar south is to find the time halfway between sunrise and sunset (listed in most newspapers). A stick placed in the ground vertically will cast a shadow that is due north-south at the true solar noontime, or ask a local surveyor how many degrees to the east or west of compass south is true south.

12. Many states are considering laws to guarantee solar access. It is important to be aware of these laws because they can both protect you and your house, or keep you from building the house you want if you will block your neighbors' solar access.

13. Trees are a major design tool for energy efficiency. They can channel winds, block the sun, admit the sun, provide evaporative cooling, and can cool an entire neighborhood by shading the ground. If they are positioned incorrectly, however, they can cool you in the winter and heat you in the summer. See footnote #7. Small, young trees can rapidly become large and block the sun. Deciduous trees with open branching patterns are the best choice for south side vegetation since they will block the sun in the winter less than any other tree. The worst trees for use with solar systems are any of the evergreens. They also happen to be among the fastest growing. Restrict the use of evergreen trees to the east and west sides of the house.

14. Winter storms generally come from predictable directions. You will want to orient your house in relationship to shading and trees so that the winter winds won't attack your house directly.

15. The dominant view is not necessarily the preferred location for solar heating needs. A spectacular view to the north might inspire frostbite instead of wonder. Views are only skin deep, so you may have to compromise your floor to ceiling glass on the north and reduce it to a smaller viewing area. Perhaps you can use a north facing view window with quilted window coverings for the night and place a solar greenhouse on the south side.

16. Ground covers are the biological reflectors of solar design. Different types of ground cover will yield different results in front of south glazing. Snow will reflect up to 90% of the sunlight striking it, while asphalt driveways will absorb 80%. Be aware in your design planning of the effect of different ground covers.

17. Many solar designs rely on large areas of south facing glass for their operation. If your south face is looking onto a traffic area or pedestrian area you may lose a lot of privacy with that particular design. Keep in mind the privacy requirements you have and balance that with some of the alternative methods of using the sunlight, e.g. clerestories, Trombe walls, and greenhouses.

18. Mass is a necessary component of a solar system. It is used to retain heat or cold and may be masonry, water, or phase change salts. It is undesirable in seismic zones to use high amounts of mass in the upper parts of the structure, so certain types of systems are more impractical. If you want a two-story Trombe wall, look elsewhere for land than California.

Additional Information

The Solar Site Selector

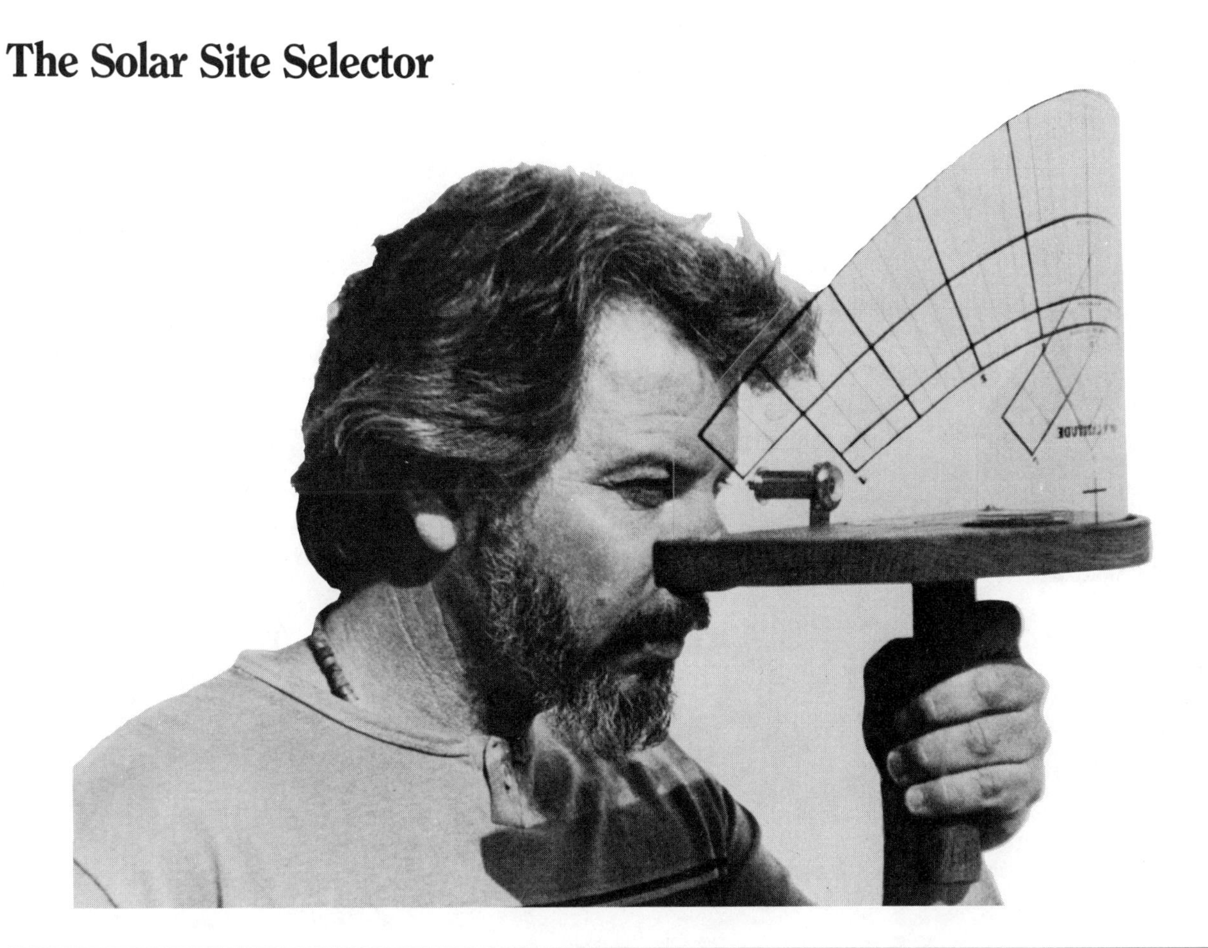

The Solar Site Selector calculates solar access and demonstrates shading patterns for any given site or surface throughout the year. This is done by using silk-screened grids that function as solar windows. When viewed through the 180-degree eyepiece, the sun paths, hour lines, and insolation segments are superimposed on the site being studied, instantly depicting shading patterns and making possible a simple, visual calculation of hours of solar access. The instrument also enables a quick computation of the percent of occlusion and percent of incident solar radiation (insolation) for that site. The instrument fits a handle (included) or photographic tripods and is oriented with a built-in compensating compass and bubble level. Take the solar site selector to the site or building you wish to study for solar access, siting and orientation. Accurate readings can be made for the entire year at any time of the day, in clear or cloudy weather.

The illustration on page 36 simulates the type of shading objects that may be encountered as viewed through the Solar Site Selector. All objects that extend above the winter solstice line will cast a shadow on the site at the time of day indicated by the hour lines.

As seen through the winter grid, the deciduous trees (shown here as viewed in the summer with full foliage) will have shed their leaves and be casting a partial twiggy shadow from 8 AM to 11 AM throughout the winter months. The house will be casting a permanent solid shadow on the site from approximately 1:30 PM through the rest of the afternoon during the winter months. There are no obstructions going beyond the spring-fall equinox line into the summer months. However, in other applications, sighting from the summer grid may be desirable.

In this illustration there are two-and-a-third hours of incident solar radiation plus three hours of partial radiation available at the winter solstice. Over five hours are available during both spring and fall equinoxes. Unobstructed radiation is available during the entire summer.

Without the elimination of at least the largest deciduous tree, the site has minimal solar access.

A simulated view through the Solar Site Selector...

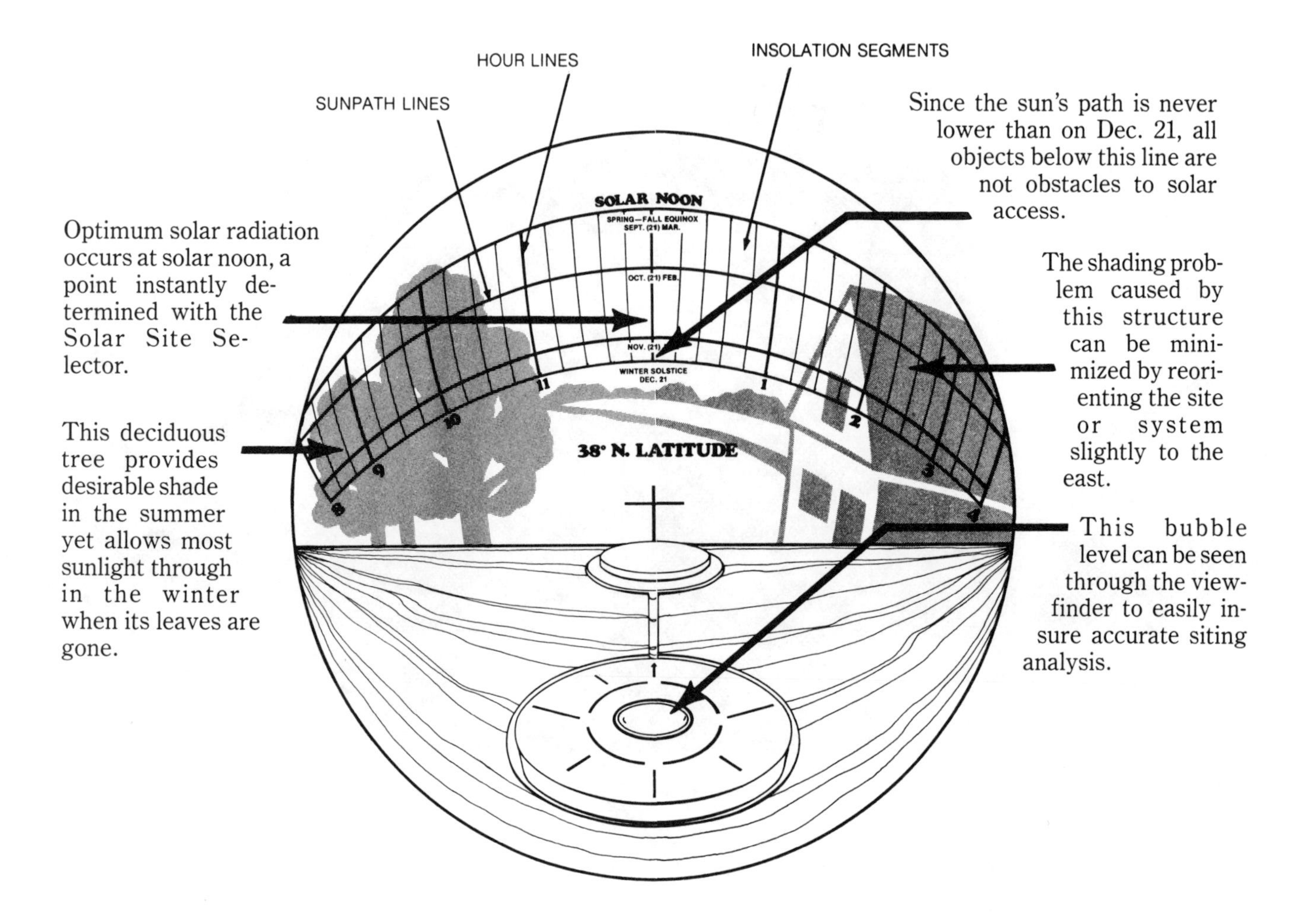

To be most efficient and cost-effective, solar glazing should have four to six hours per day of unobstructed sunshine at the winter solstice. Since much more incident solar radiation is absorbed between the hours of 10 AM and 2 PM (as compared to other hours), it is crucial to avoid shading patterns in these four mid-day hours.

Possible future obstructions to solar access should also be considered, such as immature trees not yet reaching the winter solstice line or possible future construction to the south.

For more information on The Solar Site Selector, contact:

Solar Site Selector
The Owner-Builder Center
1824 Fourth Street
Berkeley, CA 94710

To order, enclose $79.95 plus $2.00 handling (California residents add 6½% Sales Tax).

Inventory of Site

How much sunlight will fall on the site during different times of the year and different times of the day? Use solar viewer (where possible).

	9 AM	12 Noon	3 PM	5 PM
SOUTH WALL				
JUNE				
SEPT.				
DEC.				
MAR.				
NORTH WALL				
JUNE				
SEPT.				
DEC.				
MAR.				
EAST WALL				
JUNE				
SEPT.				
DEC.				
MAR.				
WEST WALL				
JUNE				
SEPT.				
DEC.				
MAR.				

A Brief Outline of Passive and Hybrid Cooling

by Don Elmer

COOLING IS A matter of human comfort and involves air temperature, humidity, air flow, the radiant temperature of surfaces, clothing, and activity levels. Unfavorable climate—high air temperatures and high levels of humidity caused by high levels of solar radiation—is a major source of human discomfort. The operation of appliances inside of buildings contributes to discomfort. Kitchens, laundry rooms, bathrooms, and various industrial and commercial processes also add heat and humidity.

Proper building design and operation are the keys to achieving comfort in overheated climates. Traditional cultures have all developed effective building adaptations to local climate problems. The tropical building which has a roof, but no walls, is an adaption to climate conditions where the air temperature is comfortable, as long as the breeze is blowing, but people get too hot in direct sunlight. The adobe pueblos, with their narrow, small windows and thick walls, are an adaption to desert conditions where the daytime air temperatures are too high for comfort, but where the nighttime coolness can be stored in the thick walls through the following day.

The owner-builder should start where all traditional cultures do with their building design—heat gain control. The first and most important part of passive cooling is preventing heat and humidity from getting into the building, and reducing or eliminating internal sources of heat and humidity.

The main exterior sources of heat are sunlight and hot air. The main techniques for reducing the effect of sunlight are reflection and shading. All exposed surfaces in overheated climates should be a light color so that they can reflect direct sunlight, and also diffuse and reflected sunlight. Roofs are especially critical, and a traditional brown shingle roof is a real problem in hot climates. All exposed surfaces should be shaded, where practical. This should include at least the east and west walls and especially the windows on these and the south wall. Windows should be shaded by overhangs, awnings, and vertical external curtains, vine covered trellises, trees, bushes, or walls. These two ways of preventing solar heat from getting into a house cannot be over emphasized. The deep wraparound porch characteristic of the South is one example of a shading technique that deserves imitation by owner-builders in hot climates everywhere.

There are some other important, somewhat more complex, ways of limiting the effect of solar radiation. High temperatures on roofs are easily reduced by evaporation. Roof sprinkler systems are common on commercial buildings throughout the Southeast. Such a system does not need to use much water, only enough to wet the roof so that evaporation can cool it. Another approach is the ice house or double roof design—a roof underneath the normal roof makes an air channel through which outside air can flow and strip the heat from the bottom of the exterior roof. This heat is carried directly outside instead of being allowed to build up in the attic where it can overheat the ceiling and make the occupants uncomfortable. Attic ventilation fans are a less effective version of this system in which the ceiling functions as the second roof layer.

The main techniques for limiting the penetration of heat from the air are insulation and proper control of ventilation. Insulation levels in hot climates have traditionally been quite low, with many older homes having no insulation at all. In order to resist heat gain, insulation levels in hot climates should be nearly as high as those for cold climates. This even includes double pane windows in extremely hot climates. Careful attention must also be given to control of infiltration and ventilation. This control is critical in cold climates so that cold air doesn't sneak into the house and make the heater run. Its equally important so that hot, humid air doesn't sneak in and make the (even more expensive to operate) air conditioner run.

Once heat gain control has been attended to, the owner-builder should consider ventilation as the first technique to use for cooling a building and its occupants. In general, air temperatures of less than 93 degrees (the approximate temperature of human skin) are useful for cooling people. Air hotter than this adds heat to the body, but may still be useful for ventilating the attic or walls since the sun may be heating them to more than 150 degrees and the air can still remove some of that heat.

Ventilation requires openings between rooms as well as in the exterior walls. Windows must be placed correctly and the rooms must permit the air to flow from one side of the house to the other. Many modern buildings are demonstrations of how not to design for ventilation since their windows do not open, or if they do, the rooms are sealed off from one another. If daytime air temperatures are too high, ventilation can still be used at night, so long as the air cools at night. This is true of most desert climates where the coolness can be stored at night in the thick walls of an adobe building, or in the thermal mass of a solar heating system. Unfortunately, this approach is not available in hot, humid climates where the air temperature remains consistently high at night.

In humid climates, ventilation is the classic solution to discomfort, but modern building design has made its application more difficult. Dehumidification becomes necessary in most building designs, at least for parts of the building. Many people find it difficult to sleep at night in muggy conditions. The owner-builder should maximize the degree to which ventilation can be used by controlling the environment only in selected spaces—home offices and bedrooms, for instance.

Ventilation air may be too hot to use both day and night. In this case there are several techniques which can be used to cool it, including the evaporative cooler and earth tubes. The evaporative cooler, or swamp cooler, has been a traditional favorite in the desert Southwest because of its low initial and operating costs. It works well wherever the air is fairly dry. New versions of the swamp cooler are being developed which will work under more humid conditions, and which will introduce less humidity into the house. Owner-builders should give a lot of attention to this cooling approach, once they have dealt effectively with heat gain control.

The earth tube is an increasingly popular but much misunderstood approach to cooling ventilation air. Air is brought into the house through buried tubing and is cooled off by the deep soil temperatures. It is commonly believed that the temperature of the earth everywhere is about 56 degrees. This is wrong. The temperature of undisturbed earth at ten or more feet in depth at any location is approximately equal to the average annual air temperature at that site. This varies in the U.S. from the upper 40s in Maine to the lower 80s in Florida. Another factor which is frequently ignored is the poor conductivity of soil. This means that earth tubes will heat up the soil around them faster than the heat can dissipate away into the surrounding soil. During the summer they will become progressively less useful as time goes by. They can work well under some circumstances where the soil is fairly moist and cool, but the initial installation expense of excavating a trench a few hundred feet long and ten feet deep makes them a low priority for the average owner-builder.

Ventilation techniques also include all the common approaches to making air move when the wind is not blowing. This means that casablanca fans, whole house exhaust fans, and even portable fans are an important part of cooling. Anytime that the air below 93 degrees flows over the skin, the skin will lose heat to the air. Evaporation of sweat will also increase and this cools the skin even more. Fans should be placed so as to blow air over places where people stand or sit in the house. The whole house exhaust fan is designed to push the hot air out of the house and performs a different but related function.

If ventilation is insufficient once the maximum feasible degree of ventilation has been incorporated, the owner-builder must consider other cooling techniques. The major problem which ventilation will not cure is humidity. Dehumidification is probably the most difficult problem to try to solve with passive means. The only practical system which the owner-builder can rely on at this time is the regular air conditioner. Combining the air conditioner with heat gain control and limiting the number of rooms to which it is applied will permit you to get by with a much smaller unit than is normal in hot, humid climates. *Mad Dog Design and Construction* is building houses in Florida which stay quite cool and comfortable with a 1/3 ton air conditioner which is basically there to dehumidify the house air.

Ventilation, evaporative air conditioners, and dehumidifiers all attempt to modify the temperature or humidity of the air in the building. There are two important techniques for cooling the building itself. These two techniques include *earth buried* or *bermed buildings*, and *roof pond systems*.

Burying or berming a building not only isolates it from the heat gain due to sunlight shining directly on roofs and walls, but also takes advantage of the soil temperatures which are usually lower than the air temperatures in the summer. These buildings are also successful in northern climates where they insulate the building from cold winter winds, and in the desert as a means of escaping hot summer winds. They have more limited application in the humid southeast where the walls may be cool enough to cause condensation problems, or where the soil is simply too warm to make it worthwhile.

Roof pond systems are probably best known under the *Skytherm®* name, developed by Harold Hay. In this system a strong roof, with lots of load bearing walls, carries large bags of water covered by moveable insulation. During the summer, the water bags are shielded from the sun during the day, but are exposed to the cool night sky. The ceiling is in direct contact with the water bags and is cool during the day. Heat from indoors is absorbed by the water bags. This approach works well in dry to moderately dry climates. In fact, in Atascadero, California, a house that Harold Hay designed stays at almost the same temperature year round. There are other popular versions of this approach but all of them share a similar limitation—they need dry climates to work best.

The essence of passive cooling lies in an attitude towards climate and nature. The very climate conditions which create human discomfort also represent the resources with which the owner-builder can work. Traditional cultures dealt with a wide variety of climates fairly successfully long before air conditioning was invented. The owner-builder should not look for a passive substitute for the air conditioner, but rather for an intuitive and informed adaption of the design to the local climate. This is an approach which requires careful attention to the facts of nature, but which can liberate you from demeaning dependency on the utility company.

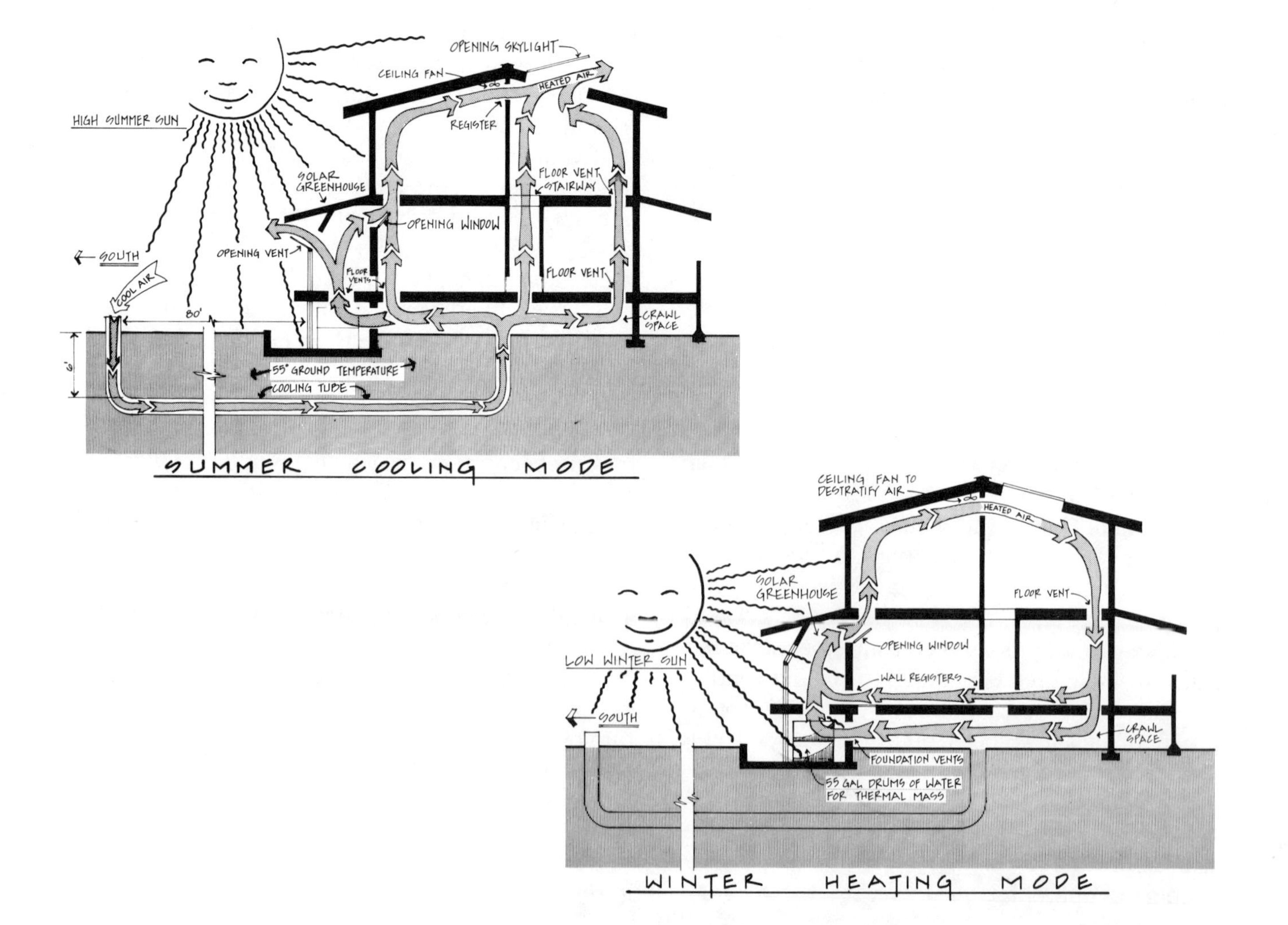

Climate Analysis: Solar Heating and Cooling

What are the maximum, average, and minimum air temperatures for both day and night during the summer months?[1]

Day	**Night**
Maximum: ____________	Maximum: ____________
Average: ____________	Average: ____________
Minimum: ____________	Minimum: ____________

What are the wet bulb temperatures during the summer?[2]

Maximum: ____________

Average: ____________

Minimum: ____________

What is the level of solar input and what are the cloudiness and clearness factors for the region?[3]

Solar input ____________

Cloudiness ____________

Clearness ____________

What are the prevailing seasonal patterns of climate and weather?

____

____

____

What are the peak weather events, such as storms, when during the year do they occur, and how long do they last?[4]

____

____

____

What are the prevailing wind directions?[5]

Summer:	**Winter:**
Day ____________	Day ____________
Night ____________	Night ____________

What are the major features of the nearby topography?[6]

____

____

____

What is the predominant nearby vegetation?[7]

____

____

____

Are any buildings near enough to the site to affect it? ☐
Are any planned for the future which might affect it?[8] ☐

What weather patterns characterize the area near the site?[9]

____

____

____

What are the dominant directions and speed for winds near the site?[10]

Direction ______________________________

Speed ______________________________

Are there any nearby bodies of water of any size?[11] ☐ YES ☐ NO

Are there any sources of air pollution near the site?[12] ☐ YES ☐ NO

What are the development patterns of the nearby area?[13]

____

____

____

What are the major topographic features of the site? Which direction(s) does the site slope?[14]

____

____

____

What are the dominant forms of vegetation on the site?

____

____

____

Where are the major deciduous and evergreen trees located on the site?

Location of deciduous trees_______________________________

Location of evergreen trees_______________________________

Are any of them to be (re)moved?[15] □ YES □ NO

What are the dominant wind flows over the site?[16]

____

____

____

What are the characteristics of the soil on the site?[17]

____

____

____

What sorts of easements exist concerning the site? Are there any solar rights belonging to neighbors which might affect the site?[18]

____

____

____

Notes

1. The designer needs to know the patterns of air temperature in order to understand the climate's problems and opportunities. Knowing how high the temperatures are is important in calculating the amount of insulation in the walls and roof, in deciding if the evaporative air conditioner can cool the outside air down into the comfort zone, in deciding if the air is cool enough to ventilate with, and in calculating the total heat load pushed into the house by the climate.

Designers generally work with average values for air temperature. These values can be found in the reports of the National Oceanic and Atmospheric Administration entitled *Local Climatological Data*. These particular reports must be a part of the basic data collecting which every home designer or builder develops. They contain a narrative summary of the climate and lists of the normal, mean, and extreme conditions for air temperature, heating and cooling degree days, rainfall, relative humidity, wind, cloud cover, and air pressure.

If the daytime temperatures are too hot for ventilation and none of the devices are cost effective, the best tactic is to close the house up during the day and cool it off at night enough to coast through the following day. If the air temperature at night is low enough it can be used to ventilate the house and cool it off for the following day. In the right climates, and with the correct design this technique can handle the cooling problems all by itself. If the air is not cool enough, then it can usually be cooled down enough with an evaporative or mechanical air conditioner.

The coolness must be preserved during the day. Closing up the house is part of the solution, and thermal mass is the other. For a house to successfully use this strategy it must include masonary walls, double gypsum board, or waterwalls which can be insulated from the exterior; or some sort of thermal storage system like a rock bed. The same calculations which are used to size passive heating storage systems can be used to size cooling storage.

2. The wet bulb temperature is measured with a special thermometer which has a wetted cloth wrapped around the fluid reservoir. It measures the air temperature as modified by the cooling due to evaporation from the wetted cloth. This is directly related to the humidity, so the wet bulb temperature is an engineering measurement of the combined influence of air temperature and humidity.

The wet bulb temperature is very important in determining if evaporative air conditioners will work satisfactorily. It also tells you if the climate will be uncomfortable due to high humidity which occurs at the same time as moderate to high temperatures. Wet bulb values in excess of about 80 degrees will be uncomfortable unless the air flow is rapid or the activity levels are kept quite low.

The evaporative air conditioner will typically bring the temperature of air passing through it down to within about 15 degrees of the wet bulb temperature. If the wet bulb temperature is at or below 70 degrees then the direct evaporative cooler will work nicely. If the wet bulb temperatures are above 75 degrees, then the system will not provide comfortable cooling under the hottest conditions.

Wet bulb temperatures are not listed in the *Local Climatological Data*. They can be found by consulting a nearby public or college engineering library for a book called the *Cooling and Heating Load Calculation Manual*. This book is published by the American Society of Heating, Refrigerating and Air Conditioning Engineers as ASHRAE GRP-158. The second chapter lists weather data and design conditions for the United States. It lists the design dry and wet bulb temperatures which should be used to calculate the design of heating and cooling systems.

3. Total solar input varies around the world according to the location and season of the year. Total solar input is important in deciding what type of heating or cooling system will function best for you. Solar input available at the surface of the earth increases as you get closer to the equator. Regional averages will be modified greatly due to the local microclimate. The San Francisco Bay area is much cloudier than would be expected from its location on the regional map, due to the tremendous influence of the Pacific and the Bay.

You should consult the *Climate Atlas of the United States* at your local library. It has maps of temperatures, rainfall, sunshine, relative humidity, and other important factors which you should consider in selecting a region in which to purchase a lot. If the library doesn't have a copy, you can get one for $6 from the National Climate Center, Federal Building, Asheville, NC 28801. They are also the source for the *Local Climate Summary* mentioned earlier.

The importance of the available sunlight is obvious in the calculation of solar heating systems. *The Passive Solar Energy Book* by Ed Mazaria, and other design books will show you how to calculate the size of your passive heating system according to the available sunshine.

The importance of available sunshine in passive cooling is not so clear. You must remember that the sources of heat which make a house uncomfortable include not only the external air temperature, but also the sun shining on the roof and walls of the building. Engineers speak of the combined effects of these two heat sources as the Solaire Temperature. What it means is that it becomes progressively more important to invest in shading for the roof and walls as the Solaire Temperature increases.

You can use this information in selecting a site without ever calculating the Solaire Temperature by observing the degree to which the site is in an area which is overheated in the summer due to high air temperatures and at the same time has high levels of sunshine. This may seem obvious, but it is a very powerful idea. The tropical coastal climates of the Gulf Coast need lots of protection from the environment not just because the air is hot, because the deserts are hotter; but because the amount of sunshine is also very high.

4. Every climate has characteristic patterns of weather across the seasons of the year. Some of these patterns are surprising to owner-builders who are trying to select or evaluate their prospective building sites. It is very rare to find a place in the U.S. which has the classic four seasons evenly spread about the year.

Some of the surprises can be unpleasant. In the desert Southwest evaporative coolers are popular because they are inexpensive to buy and operate and because they work well in dry climates. Unfortunately the climate does not stay perfectly dry all summer long. In the months of August and September there is a mini-monsoon season during which the moist winds from the Gulf of Mexico penetrate far into the desert. During these periods of high relative humidity, the evaporative air conditioners fail to provide cool enough air for comfort. As a result they have fallen out of widespread use in many cities in the Southwest.

The hot desert winds of Southern California are another unpleasant surprise which prevent ventilation from being a useful cooling approach throughout the summer. The San Joaquin Valley has summer periods when the humidity rises dramatically at the same time that air temperatures remain high both day and night. This fact interferes with the ability to cool buildings by ventilating them with cool nighttime air, which is otherwise an effective solution for that climate.

It is important to study the narrative section of the *Local Climate Summary* to identify these patterns.

5. Regional prevailing wind patterns are important in selecting sites. In some areas, winds are totally unpredictable. In these cases choosing the site for natural ventilation will require a flat site without obstructions in any direction. For areas with predictable wind directions, such as tropical islands like Puerto Rico, it is possible to select a site with exposure to the trade winds in order to maximize the amount of natural ventilation available for cooling. The same strategy applies in passive heating conditions when the site is chosen with obstructions to the north or northwest in order to minimize the heat lost to the cold north winds of winter.

Prevailing wind directions are summed up in both the *Local Climatological Data* and in the *Climate Atlas of the United States*. The information is expressed in the form of 'wind roses' or arrows with numbers attached. Use this information as an overall preliminary summary of wind conditions, but talk to the local weather service and learn how the regional and local conditions differ. Finally, if the use of natural ventilation is an important strategy for your house, observe the actual wind patterns at your site or ask someone who has lived nearby and observed these conditions. You can expect the site conditions to differ greatly from the regional averages. Wind is very capricious.

6. Nearby topography is important in site planning for both solar heating and cooling.

In the case of the solar heating system, whether it is active or passive and whether it is for space heating or water heating, the main concern is that the collecting surface in the system should *not* be shaded during any significant part of the winter season. This means that those of you with building lots in valleys need to do some plotting of the seasonal sun angles to be sure that the hills on either side don't unnecessarily shade your collecting surface in winter.

In the case of the solar cooling system, the main concerns are more complex. The designer is concerned about finding sources of shading which might shelter the building from early morning or late afternoon sun. They are also concerned about wind and possible nearby obstructions to wind flow, like hills or forests. For cooling purposes you want winds to flow over your site if they are reasonably cool, but hot dry desert winds should be redirected away from the site by nearby heights.

Unfortunately it is hard to balance the need for solar input for heating with the need to exclude it for cooling if you are depending on nearby hills or mountains to shade the system at the right time. The winter sun is low in the sky and in the morning and evening the lot will be shaded earlier than necessary if there are hills or mountains close to the site. The summer sun is much higher so the relative effect of the nearby heights is lessened. Such compromises are associated with many mountain sites.

7. Nearby vegetation is important for reflecting solar radiation and the wind.

The more important reason for caring about nearby vegetation is that this vegetation plays an important role in determining the temperature and humidity levels of winds which blow across them on their way to the site. Desert areas typically heat up and dry out regional winds which may start out at the nearby ocean quite cool and moist. Forests may both cool and humidify dry hot winds. Crops add many tons of moisture to hot dry winds which blow over them.

The effect of vegetation on cold winter winds is much smaller since evapotranspiration of water from plant leaves is at a minimum in the winter and it is this evaporation which cools and moistens hot dry winds.

Note the dominant vegetation and use your senses, supplemented by a cheap thermometer, to see if the vegetation has any perceptible effect on the wind; and use your eyes to determine if the site receives reflections from nearby areas in either the summer or winter. There is no substitute for an intimate knowledge of your building site based on direct observation during all four seasons of the year.

8. Nearby buildings can have several important impacts on the site—shading, reflecting light onto the site, obstructing breezes, and heating up breezes.

Obviously the larger and taller the buildings are, the more impact they will have on shading the site. Large buildings will also affect the wind patterns and temperature. Buildings with reflective glass can dramatically increase cooling problems for homes in hot climates. This is a real problem throughout the southwest and south with large office buildings on the boundaries of, or integrated into, residential neighborhoods.

You can avoid obvious problems here by observation of the shadow or reflected light patch from nearby large buildings. You can even imagine the effect of the excess heat from the air conditioners of these buildings. But it is much harder to plan for the effect of the large buildings which may be built nearby in the future. There are no negative solar rights yet to protect you from trespass by reflected light from the nearby bank tower and no legal consideration given to the loss or heating up of breezes due to nearby large structures.

If you are building in an urban or suburban area, at least be sensitive to the problems which could arise if you build next to a large empty lot zoned C.

9. The main differences in regional climates are the result of topography, vegetation, proximity to bodies of water, cities, and sources of pollution.

Regional climate tells the designer about general climatic conditions, but cannot reveal the critically important differences in the weather patterns in the area near the actual site. An example of this is the well documented fact that rainfall downwind from major shopping malls in Houston, Texas, is several inches per year higher than the city wide average because the heat from the shopping mall's buildings and (solar heated) parking lots causes the moist air to rise up high enough so that the air cools off and the moisture condenses out as rain.

Other factors which affect regional climate include the existence of hills or mountains which redirect dominant wind patterns; croplands which cool and humidify winds; changes in elevations which place sites up into prevailing winds; bodies of water which can heat up or cool off winds but usually add humidity to the winds; cities which invariably heat up winds; and sources of air pollution which obscure the sunlight and increase rainfall.

The designer needs information on the relative importance of these effects in order to know if the weather conditions near the site are very different from the usual regional conditions. If any of these conditions are important, the design can respond to that fact. It's much harder to adapt a building after its been built.

10. The designer is concerned with the major patterns which occur on a seasonal basis. The direction of the wind will tell the designer what to look for. Winds off the desert will usually be hot and dry, those off the ocean will usually be hot or cold and wet.

11. The importance of nearby bodies of water bigger than the farmer's trout pond is considerable. Even smallish lakes can dramatically cool hot summer winds as they blow over them. Oceans, obviously, are even more important and have a major moderating effect on all coastal regions of the country, tending to stabilize temperature and humidity extremes.

Bodies of water are also an important source of reflected sunlight and glare. They can add to solar input to a building in both summer and winter, but the consequences of this are usually not large unless the building site is right on the water or on a slope overlooking the water. The added sunlight can be a blessing in winter and a problem in summer.

12. Air pollution is important for many reasons. It can make the ventilation approach to cooling useless. All large buildings include an air washing section in their air conditioning systems which sprays water in the incoming ventilation air and washes the filth out of it—mostly. Homes can rarely afford this luxury. The evaporative air conditioner, however, does air washing at the same time that it cools.

Air pollution also obscures the sunlight and attacks the materials in a solar system and building over the long term—not to mention its deleterious effects on vegetation and people. If you have a choice, don't build where urban pollution will interfere with your ability to open the windows when you want; or where it's going to kill the shade trees or affect your family.

13. Many suburban, and semi-rural sites are rural when they are first acquired. Far too often the owner-builder wakes up one morning to find that the city has moved in on every side. This *will* affect the comfort characteristics of your home and the performance of your solar system. Cities will heat up and dirty the air. They will block cooling winds and reflect excess light onto your site. They will also destroy the last of your aesthetic easement on the nearby area which probably played a major part in selection of your site in the first place. Ponder this when you are selecting your lot. There are no guarantees, but forethought here might save you heartache ten or twenty years from now.

14. The major topographical features of interest to the designer are the overall slope of the site, and the existence (or not) of any elevations on an otherwise level site. Sloping building sites offer the opportunity to take advantage of seasonal temperature differences between the bottom of a valley and the top of the ridge along the valley. The valley bottom is hotter in summer and colder in winter. These differences can be observed in the positioning of pioneer's homes in every part of the country, where many more homes are found on the slopes than on the valley floor—except in areas with mild seasonal differences.

A sloping site also offers a choice of an unobstructed winter location for the building which maximizes the available sunlight. South facing slopes are obviously more useful than north facing slopes for climates with wintertime heating requirements. In overheated climates the north slope can be much superior to the south slope because it shades the house from the sun in the early morning and late afternoon.

The slope also offers a chance to put the building up into the prevailing breezes. As a general rule, wind speed increases semi-logarithmically with increasing height; or, more simply, the higher the house is placed on the slope, the faster and more regularly the wind will blow.

Sloping sites also offer the opportunity to have an earth tempered, bermed, or buried house at slight additional cost. A building site on a slope must be either leveled or the building must be put on piers. Leveling usually means digging into the slope, which makes a niche for the building and provides the excess soil for berming. The advantages of earth tempering are that the exterior of the building does not face the harsh temperatures of either wintertime or summertime air. The soil acts as both insulation and thermal mass. Furthermore, for those climates with cooling problems, a bermed or earth tempered house offers the possibility for using vegetation for evaporative cooling of the soil around and on top of the house. Keep this in mind when selecting your site, if you like earth tempered housing.

15. An ideal site has no vegetation to obstruct the winter sun and plenty of vegetation to obstruct the summer sun. This design problem was solved long ago by nature with the evolution of deciduous trees which shed their leaves in the winter and leaf out in spring.

At least this is the ideal case as portrayed in innumerable solar books. Alas, things are never this way. Large trees can still obscure 50% of the sunlight with their bare branches in wintertime, as many a builder of an add-on greenhouse in New England has discovered and had to face chopping down or trimming a favorite tree which was to the south of the greenhouse. Also, the tree may not drop its leaves as soon as you would like.

In southern climates, deciduous trees may reverse their inherited seasonal pattern and keep their leaves all winter, and shed them in the hot dry summer because of the heat and lack of water. Some deciduous species hve adapted to this exact pattern and many northern imports in southern cities forget their origins promptly. Unfortunately this is just the time of year when their shade would be useful to avoid summertime overheating. A frequently overlooked solution is to provide significant amounts of water to these trees so that they will keep their leaves through the summer.

Shade is not the only desirable attribute of trees. Trees and bushes have always been used to shelter houses from harsh winter winds and to channel summer breezes into the house in warmer climes. This aspect of site vegetation can be as important as shading, and permits the designer to use less expensive bushes and shrubs to have an impact on the house's comfort conditions.

In general, plant trees or bushes on the north side (at least in the northern hemisphere and north of about 20 degrees north latitude) in order to shelter the house in winter. Use evergreen species which will present the largest mass to resist the wind. Plant bushes and trees on the other sides to funnel prevailing cool breezes into the house in the summer and spring. These can be either evergreen or deciduous.

Use the existing site vegetation to maximum advantage when you place the building on the site and orient it. Also use the existing vegetation as an indicator of what will grow well at your site.

One frequently overlooked function of site vegetation is its effect on the radiant temperature of the surroundings of the house. An asphalt parking lot which heats up under the summer sun will radiate a significant amount of excess heat onto a building. A grass lawn or a forested area will remain much cooler due to evaporation and will not add to the building's heat load through infra-red radiation. So do not become overzealous in pruning back the jungle you find on your lot when you buy it. Balance the need for sunlight in winter with the desire to avoid it in summer.

16. The designer needs to know if the site is subject to wind flow, and—if so—which direction it comes from. This knowledge will enable the designer to place the house and orient it in such a way as to use the site's topography and vegetation to avoid winter winds and capture summer winds—if they are cool enough to be useful. Often the difference from one corner of a site to another is enough to place the house where it can have access to cool summer breezes.

Knowledge of the prevailing wind patterns should also determine the landscaping plan. The purpose of this plan is to compliment the site's natural advantages or to alleviate its disadvantages with respect to winter winds and summer breezes, as well as sunlight.

17. The soil characteristics are important in planning for earth tempered, bermed, or buried houses. The main features which matter are its structural properties as a building element, its permeability by water, and what underlies it. Excavation costs rise in direct proportion to the difficulty in digging holes. This is determined by the structural properties of the soil. Some clays are so incompetent that complex and expensive shoring and bracing must be used during construction. Some clays expand so much during rain storms that they crush supports and forms.

Sites which have a rocky layer just under the soil are not ideal candidates for buried buildings. However, bermed buildings can be built on solid rock if you can afford to truck in the soil to berm with. Just make sure that you have some idea of the soil conditions and presence or absence of rock before spending months planning the perfect troglodyte home.

Soil permeability by water is important for those planning to use earth tubes. The temperature of the soil at ten feet of depth is usually about equal to the average annual air temperature. However, major deviations can occur if the soil is very dry. Soil's ability to conduct heat is also a function of soil moisture and it is necessary to keep the soil around earth tubes as moist as possible if you want to cool with them throughout the summer. Obviously, hard clay is more difficult to keep moist than sand, and granite is nearly impossible. Think about these facts before planning an earth tube system. Remember also that the soil is a relatively poor conductor of heat, and the use of earth tubes will heat up the soil and make it progressively less able to absorb summer heat from ventilation air.

18. Easements across your site can be owned by neighboring land owners, governments, and government licensed monopolies—otherwise known as utilities. Any of these parties could come on your site at some point and cut a swath of trees, or clear a significant chunk of your lot for a county road, or just trim the trees back to shrubs in order to string cable TV to the resort cabins up the valley. These sorts of adjustments in your site can have important and seldom useful impacts on the comfort of your house and the functioning of your solar system.

In this modern era, you must also do some legal sleuthing and discover whether or not your neighbors have a solar easement which places restrictions on the height, location, shape, or exterior treatment of your structure. You want to be considerate of their solar access whether such laws or provisions exist anyway, but some solar access covenants can be overzealous. It's a new field and the boiler plate language is not quite worked out yet to everyone's satisfaction.

Many cities and counties are establishing laws which set out solar access rights. These give you guaranteed access, if the law is followed and utilized, and are frequently the source of your neighbor's rights. Study your local provisions and set a good example with your design. We all want to encourage our neighbors by our lower utility bills and higher comfort levels to 'go solar' too, but without foresight we may find that we have blocked their optimum site location.

Additional Information

Additional Information

3/Water Systems

Knowing that our bodies are mostly water, the need for good clean water is obvious.

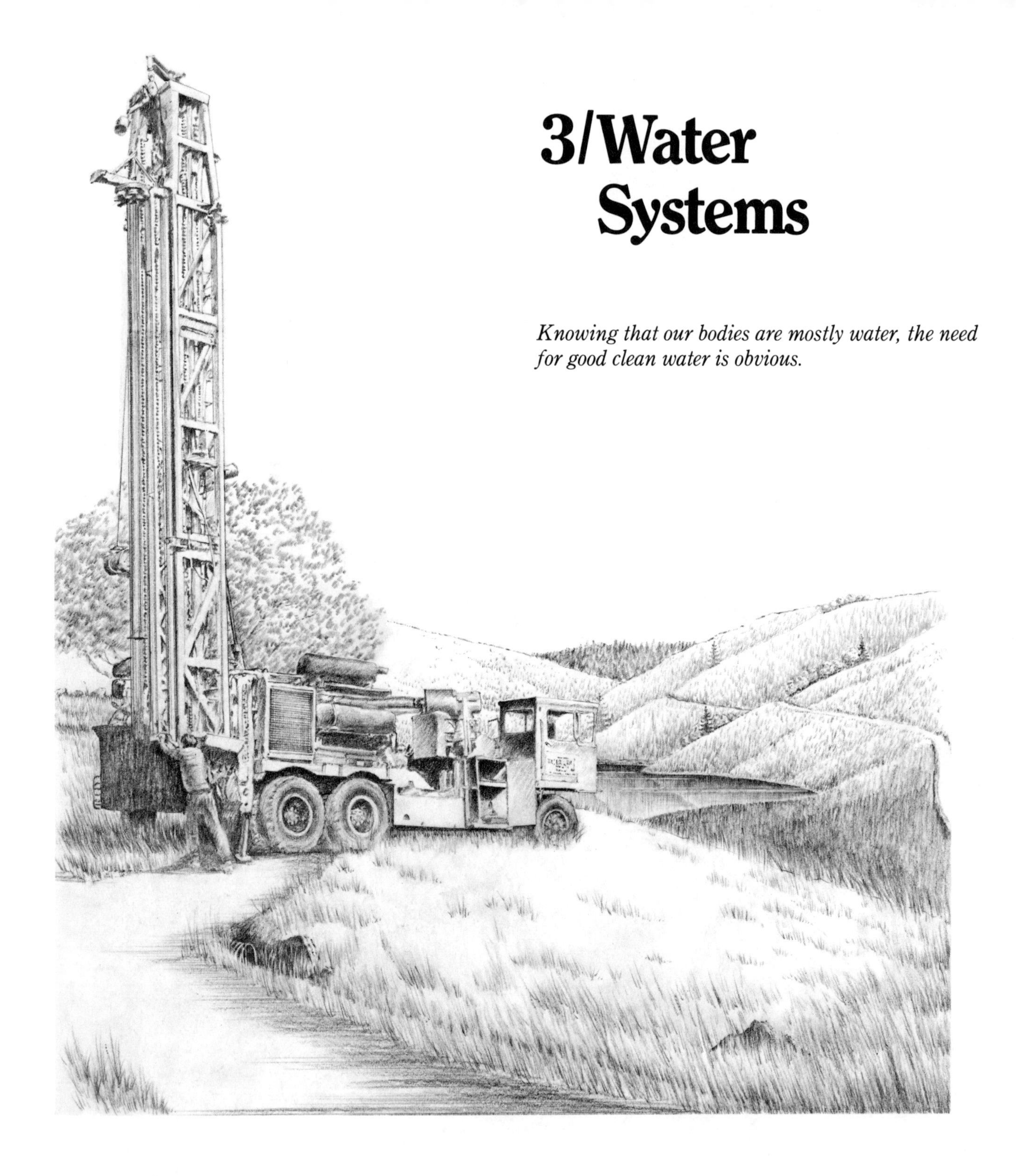

WHETHER YOU ARE digging a well on a piece of rural property or hooking into an existing system in a subdivision, there are things you need to know about the proposed or existing water system. Hooking into existing water systems is a good bit less complicated than drilling your own well, but sometimes even this can be troublesome. In either case the need for a constant, dependable supply of water is necessary in the success of your project.

Unfortunately the supply of pure drinking water in this country is diminishing. The EPA states that only 10% of our creeks and streams are unpolluted, and every day we hear more about commercial, industrial, and agricultural wastes polluting underground water tables and thereby neighboring wells. There is even evidence that the chlorination of many systems may be dangerous to our health. Because of this it may be wise to investigate the purity of the water you are planning to hook into. If it is found wanting, there may be a local spring water service nearby that can supply your drinking water at a reasonable price.

This chapter has been divided into three sections: one for existing systems, one for existing wells, and one for questions having to do with digging your own well. Again, these may be questions you want answered before you purchase the land since the lack of good water or a prohibitively high cost of drilling a well can render the project unworkable.

For Further Reading

Manual of Individual Water Supply Systems. Environmental Protection Agency. Superintendent of Documents, Government Printing Office, Washington, D.C. 20402.

Planning for an Individual Water System. American Association of Vocational Instructional Material. AAVIM, Engineering Center, Athens, GA 30602.

Water Supply for Rural Areas and Small Communities. Wagner and Lanoix. World Health Organization, Q Corporation, 49 Sheridan Avenue, Albany, NY 12210.

Water Systems

In what manner will water be supplied to your home?

- □ City or county system
- □ Community or group owned system
- □ Pre-existing well
- □ Well to be dug by you
- □ Developed spring
- □ Other ______________________

Is there a stream nearby where a hydraulic ram can be used?[1] □ YES □ NO

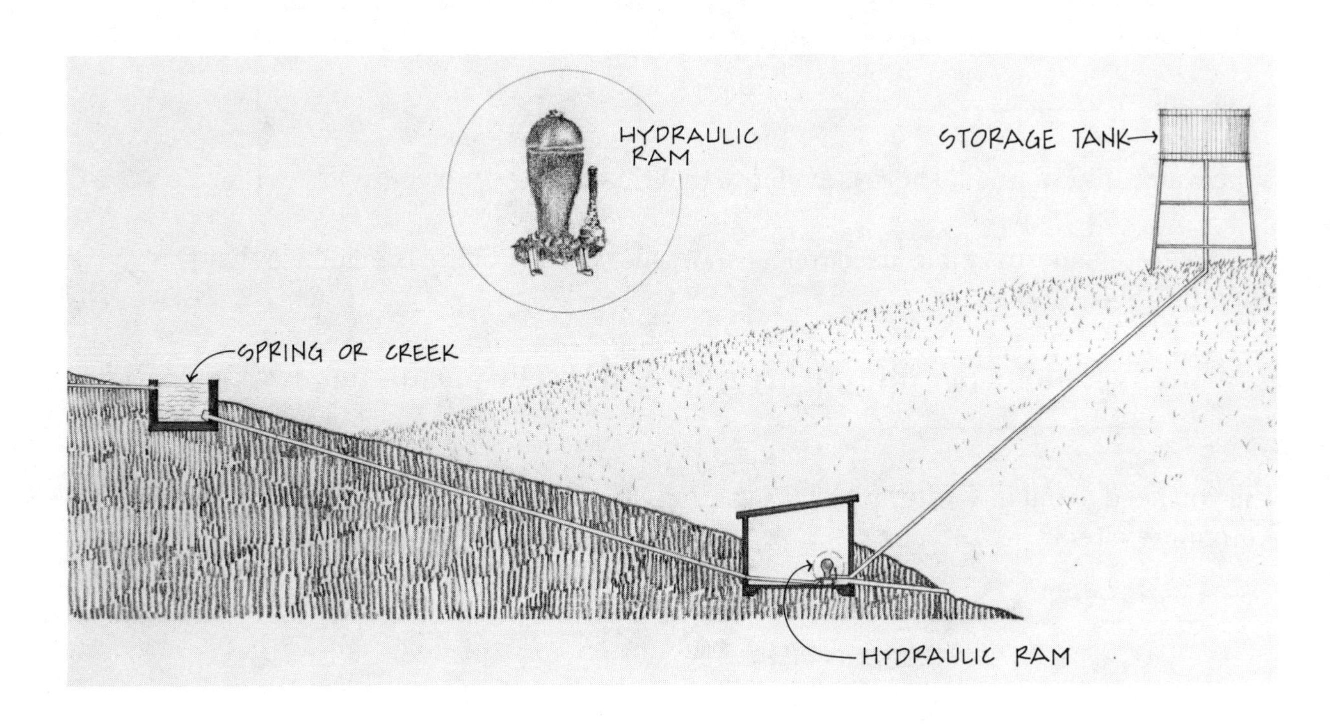

If your water is to be supplied by an existing municipal or group owned system, answer the following questions. If your water is to come from an existing well, answer the questions beginning on page 54. Finally, if you must dig your own well, answer the questions beginning on page 56.

EXISTING SYSTEMS

If your water will be supplied by a city, county, or group owned system:

How much will hookup fees cost? ____________________

How much are yearly maintenance fees? ____________________

What is the cost of water? ____________________

Are rates fairly stable? ____________________

Is there a limit on how much water can be used? ____________________

What will be the cost to run a pipe to the water source? ____________________

Is the water pressure:

- ☐ Good
- ☐ Fair
- ☐ Poor

Is there a minimum required water pressure? ____________________

A maximum? ____________________

Is the water fluoridated? (There is evidence that this could be unhealthy) ☐ YES ☐ NO

Are concrete/asbestos pipes used to transport the water? (There is evidence that this could be unhealthy) ☐ YES ☐ NO

How many gallons per minute are required by the permit agency?[2]

____________________ GPM

Will you need to put in a water filtering system? ☐ YES ☐ NO
Softening system? ☐ YES ☐ NO

How pure is the water?

- ☐ Pure
- ☐ Moderate
- ☐ Impure

What is the pH of the water? ____________________

Is there a possibility that any of the following may contaminate the well:

- ☐ Pesticides
- ☐ Industrial waste
- ☐ Dead animals
- ☐ Septic systems
- ☐ Tree spraying

Are there any water rights involved? ☐ YES ☐ NO

In existing systems, does the hookup really exist? ☐ YES ☐ NO

Where do you go to locate water hookup? ____________________
Are these locations accurately recorded? ☐ YES ☐ NO

Who is allowed to make connections to existing system? ____________________

What kind of fittings and pipes are needed for the hookup?

____

____

____

Will the water pipes need to cross any:

- ☐ Roads
- ☐ Tree roots
- ☐ Cables

Is a tracer line needed? ☐ YES ☐ NO

Will a heat tape be needed to avoid freezing? ☐ YES ☐ NO

Is there any water rationing or scheduling in the area? ☐ YES ☐ NO

Can water testing be done by a government agency? ☐ YES ☐ NO

Can the water system be used in a house fire? ☐ YES ☐ NO

Have your neighbors had any water problems? ☐ YES ☐ NO

____

____

Is your water source permanent? ☐ YES ☐ NO

EXISTING WELLS

If your water will be supplied by an existing well:

How deep is the well? ____________________

What type of casing is in use? ____________________

Is the pump in good condition? ☐ YES ☐ NO
If not, what is the replacement cost? ____________________

Is the water potable? ☐ YES ☐ NO

Do any special mineral problems with the water exist? ☐ YES ☐ NO

____

Is the water table stable in the area? ☐ YES ☐ NO

Has the well ever run dry? ☐ YES ☐ NO

How far is it from the well to the house? ____________________

How far is it from the well to the septic system? ____________________

How far is it from the well to the garden? ____________________

How much will a pressurized tank, pump, and connections cost?

Pump Co. #1: ______________________________________

Type of pump ______________________________________

Size of tank ______________________________________

Other Equipment ______________________________________

Quote ______________________________________

Pump Co. #2: ______________________

Type of pump ______________________

Size of tank ______________________

Other Equipment ______________________

Quote ______________________

How many gallons per minute is the well required to produce by the permit agency?[2]
______________ GPM

Will you need to put in a water filtering system? □ YES □ NO
Softening system? □ YES □ NO

Is there a minimum required water pressure? ______________

A maximum? ______________

How pure is the water?

- □ Pure
- □ Moderate
- □ Impure

Is there a possibility that any of the following may contaminate the well:

- □ Pesticides
- □ Industrial waste
- □ Dead animals
- □ Septic systems
- □ Tree spraying

Can water testing be done by a government agency? □ YES □ NO

What is the pH of the water? ______________

Will the water pipes need to cross any:

- □ Roads
- □ Tree roots
- □ Cables

Will a heat tape be needed to avoid freezing? □ YES □ NO

How deep will the line from well to the house have to be? ______________

Can more than one house use the same well?[3] □ YES □ NO

Can the water system be used in a house fire? □ YES □ NO

Is a wellhouse needed? □ YES □ NO

Can creek, lake, or spring water be used for gardens and lawns, and the well used only for household supply? □ YES □ NO

Have your neighbors had any water problems? □ YES □ NO

____

____

Will the storage system be uphill or downhill from the well? ____________________

From the house? ____________________

Is your water source permanent? □ YES □ NO

Are existing wells in legal locations? □ YES □ NO

If central water systems are added in your area in the future, are you required to hook up to them? □ YES □ NO

DIGGING WELLS

If a well needs to be dug:

How will you decide on location of well?[4]

- □ Witcher
- □ Geologist
- □ Well digger
- □ Fate

How deep are the neighboring wells and what are the GPM?[5]

Neighbor #1 ____________________ deep, ____________________ GPM

Neighbor #2 ____________________ deep, ____________________ GPM

Neighbor #3 ____________________ deep, ____________________ GPM

Are there many dry holes in the area? □ YES □ NO

Have your neighbors had any water problems? □ YES □ NO

____

What do geology maps say about local water table and underlying rock formations?[6]

____

____

____

Well diggers:

Well digger #1:

Name ____________________________________

Address ____________________________________

Phone ________________

Type of casing to be used ________________________

Price per foot ________________ Set-up charge ________________

Additional charges ________________

Are any guarantees offered? □ YES □ NO

If so, what? ____________________________________

What provisions are made if no water is found? ____________________

Is there additional charge for drilling through rock? ________________

When can he start? ____________________________

Well digger #2:

Name ____________________________________

Address ____________________________________

Phone ________________

Type of casing to be used ________________________

Price per foot ________________ Set-up charge ________________

Additional charges ________________

Are any guarantees offered? □ YES □ NO

If so, what? ____________________________________

What provisions are made if no water is found? ____________________

Is there additional charge for drilling through rock? ________________

When can he start? ____________________________

Will the drilling rig have easy access to drilling site? □ YES □ NO

How far is it from the well to the house? ____________________

How far is it from the well to the septic system? ____________________

How far is it from the well to the garden? ____________________

How much will a pressurized tank, pump, and connections cost?

Pump Co. #1: ____

Type of pump ____

Size of tank ____________________

Other Equipment ____

Quote ____________________

Pump Co. #2: ____

Type of pump ____

Size of tank ____________________

Other Equipment ____

Quote ____________________

Will you need to install a water filtering system? □ YES □ NO
Softening system? □ YES □ NO

Is there a minimum required water pressure? ____________________

A maximum? ____________________

How pure is the water?

□ Pure
□ Moderate
□ Impure

Is there a possibility that any of the following may contaminate the well?

□ Pesticides
□ Industrial waste
□ Dead animals
□ Septic systems
□ Tree spraying

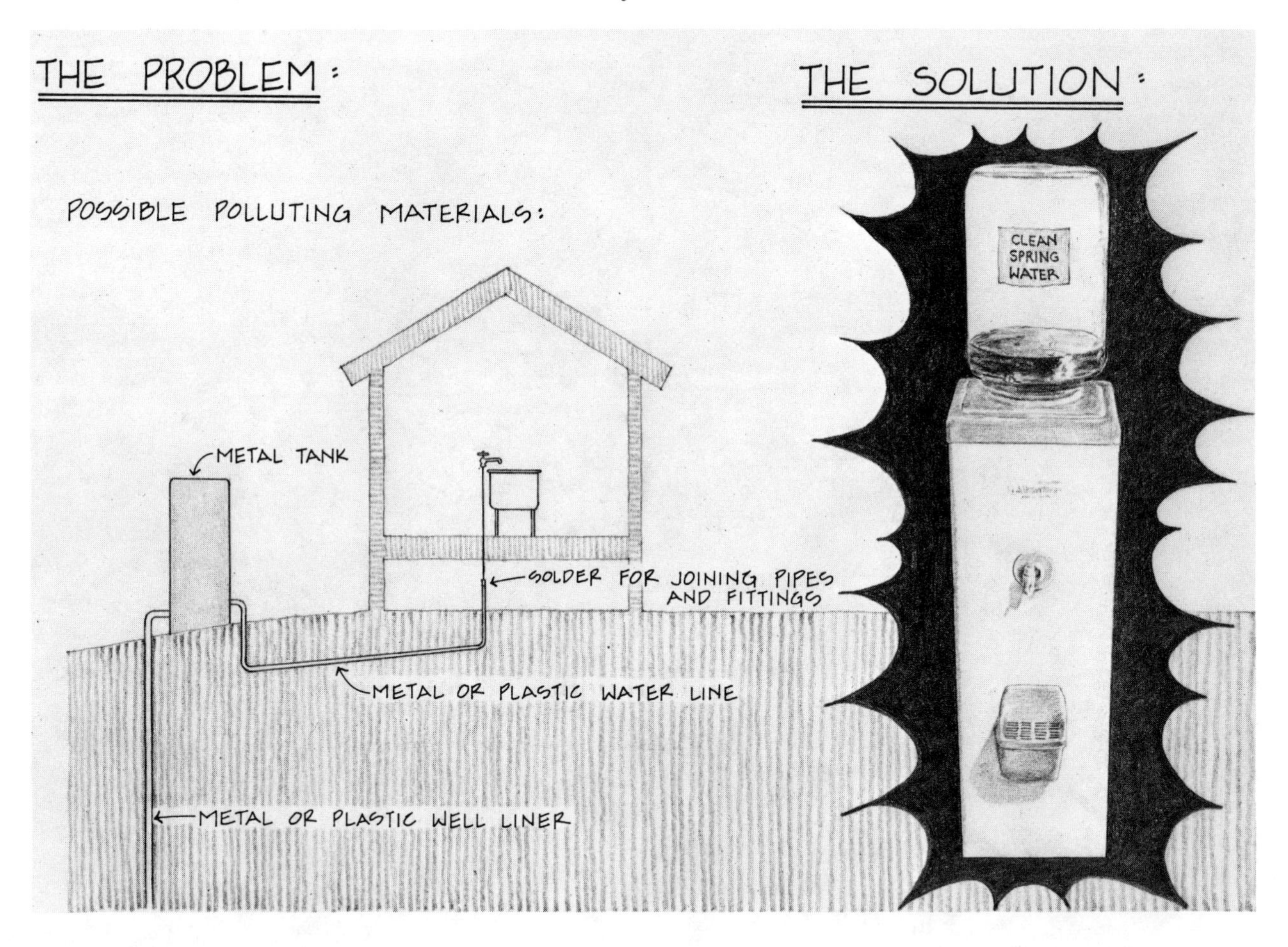

What is the pH of the water? ____________________

Can water testing be done by a government agency? □ YES □ NO

Will the water pipes need to cross any:

- □ Roads
- □ Tree roots
- □ Cables

Is a tracer line needed? □ YES □ NO

Will a heat tape be needed to avoid freezing? □ YES □ NO

How deep will the line from the well to the house have to be? ____________________

Can more than one house use the same well?[3] □ YES □ NO

Can a trench for the water line be shared with other utilities? □ YES □ NO

Can the water system be used in a house fire? □ YES □ NO

Is a wellhouse needed? □ YES □ NO

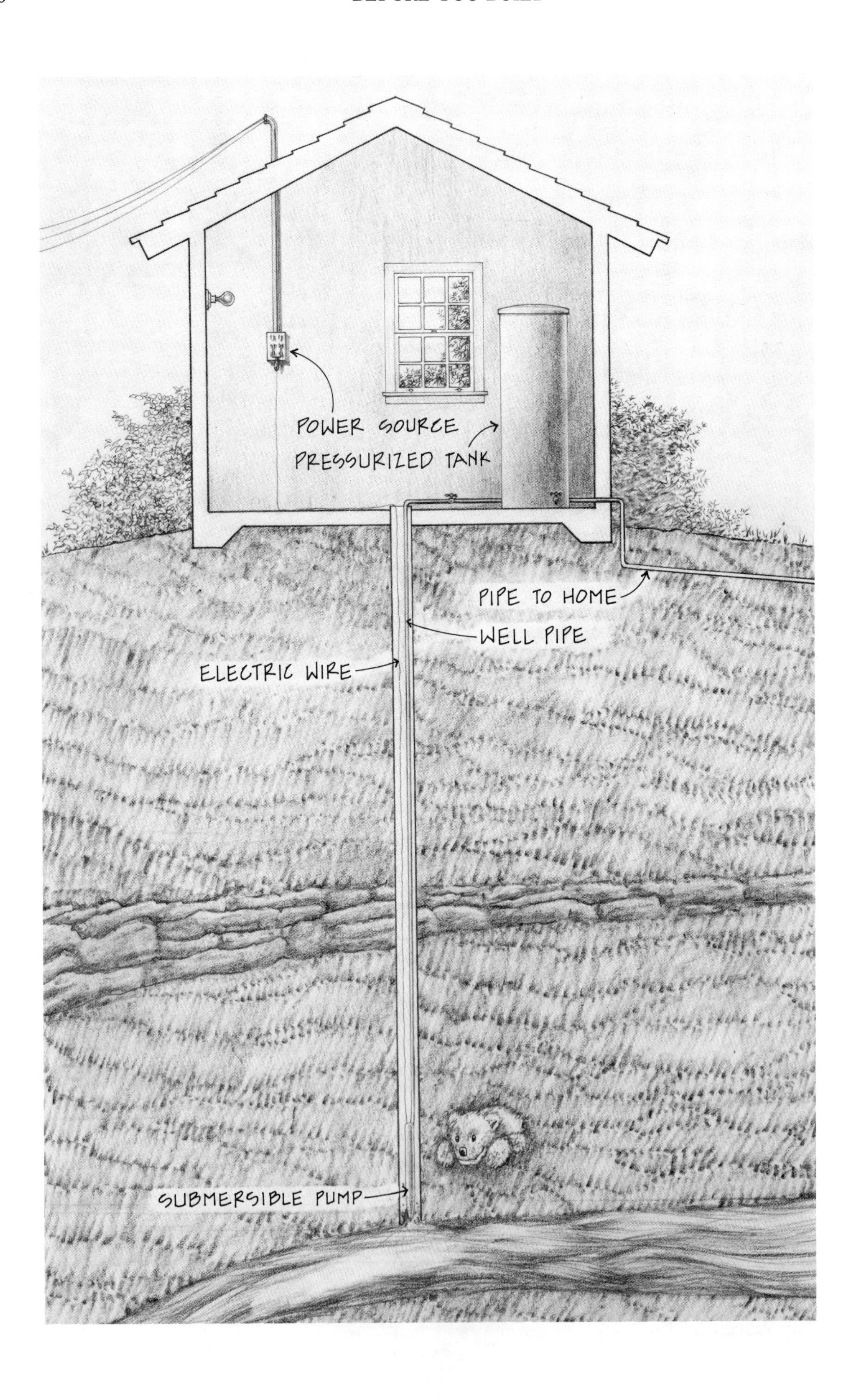
POWER SOURCE
PRESSURIZED TANK
PIPE TO HOME
WELL PIPE
ELECTRIC WIRE
SUBMERSIBLE PUMP

Can creek, lake, or spring water be used for gardens and lawns, and the well used only for household supply? □ YES □ NO

Will the storage system be uphill or downhill from the well? ____________________

From the house? ____________________

Is your water source permanent? □ YES □ NO

If central water systems are added in your area in the future, are you required to hook up to them? □ YES □ NO

Is digging your own well a possibility?[7] □ YES □ NO

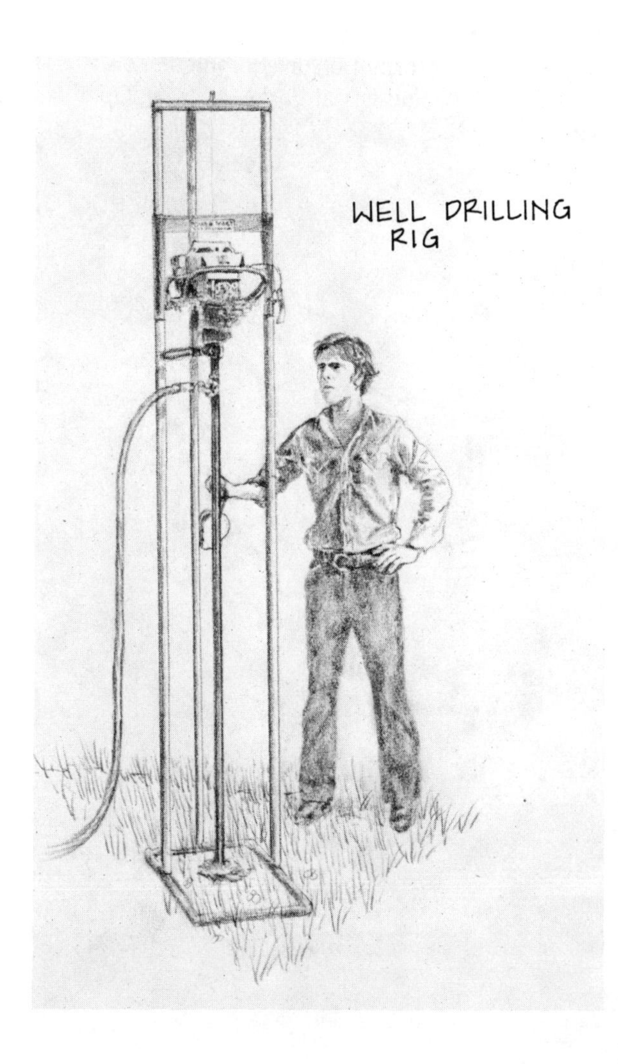

Notes

1. Hydraulic rams are mechanical devices that require no outside source of energy other than a stream or creek. The flowing action of the creek powers the ram to push water uphill, up to 100 feet if the stream is fast enough. This can then be used for the running water in your home. For more information, contact: Rife Hydraulic Engine Mfg. Co., 132 Main St., Andover, NJ 07821.

2. In areas where water is scarce or droughts long, the building department may require proof that the well can produce so many gallons per minute and therefore can be trusted not to run dry. Often, before the building permit is issued the well must be dug and the inspector will visit the property to test the well. The test involves pumping water from the well at a certain rate, say 4 GPM, and see whether it runs dry within an hour. Sometimes there are two sets of requirements—one for if you are testing in the dry season, another for the wet season.

3. Frequently, a well will supply more than enough water for one home and you may want to share the initial cost and maintenance with a neighbor. Sometimes however, if there is a lending institution involved they will often require that each home has its own water source.

4. There is much controversy on the best way to chose the location of a well. At this time there is no certain way. Some believe in geologists, others in witchers, others in fate. Well diggers have no stake in finding a way to dig the shallowest well since they are paid by the foot. I have had considerable experience in projects that involved well digging and heard of many others and I have formed a few opinions. Firstly, well witching does work and it is not magic. In ancient times it was called dowsing and was used to locate the site of cities and temples in China. It is merely a process by which some people, through the use of a certain medium or dowsing rod, locate the energy that is given off by water running through the earth. Plumbers have often used copper rods to dowse water pipes. Secondly, I have learned that an inexperienced dowser, and there are many, is able to locate areas of the land that have water flowing below but is often unable to say how many GPM and how deep, and is therefore useless in assisting you in locating your well. If you know of a dowser who has a good reputation, I would advise following his or her advice.

5. Even though knowing the depths of the neighbors' wells is of value, do not place too much weight on this information. A friend dug a well in North Carolina 130 feet deep and the church across the road had to go down 450 feet. Information about others' wells is most valuable in ascertaining if your area has problems with locating water at reasonable depths. Find out if there is any history of dry holes or very deep wells in the area and, if so, see if any of the well diggers will give guarantees of any sort.

6. Again, only take these with a grain of salt.

7. The idea of digging your own well is not so far-fetched as you might think. Shallow wells (twenty to forty feet) have been dug by hand for many years. Unfortunately, these only tap the shallow water table and often run dry and are more easily polluted. The more dependable deep wells tap underground water systems farther below the surface. These wells are usually dug by a professional with expensive equipment but there is equipment available that will allow you to do it yourself. The equipment is relatively inexpensive ($700–1,000.00) and can be resold after the well is dug. Rental well drilling equipment may also be available in your area. Thousands of dollars can be saved this way. How difficult it will be to drill your well this way will depend on how much rock you hit while digging. If you do not hit rock, you can dig up to thirty feet a day; if you hit rock, the rate may be only one foot per day. For more information on equipment, contact: Deep Rock Mfg. Co., Industrial Park, Box 870, Opelika, AL 38601.

Additional Information

Additional Information

4/Power and Phone

Electricity, to make your work easier and quicker, and a phone, to keep a link to the outside world and its resources, are two essentials in any building job. Unless one or both are absolutely unavailable, have them installed prior to the beginning of construction. The job is demanding enough without adding to it. It is best to make use of the beneficial technology at hand.

Power and phone companies often move slowly; sometimes their pace can only be matched by the government or owner-builders building their first house. Easements, engineering, and scheduling can often take months. Be sure to begin the process of obtaining electricity and telephone long before the scheduled beginning of construction and always follow up to be sure it is progressing.

Power and Phone

Is power available at the building site? □ YES □ NO

If not, how far from the building site is the closest power line? ______________

What will be the cost of bringing power to your site?[1] (obtain quote from your local power company.)

Above ground ________________

Below ground ________________

When can the power be put in?[2] ______________________________

Can this date be guaranteed? □ YES □ NO

What is the name of the person at the power company with whom you are dealing? ______________________________

Phone: ________________

How many AMP service will be required? ______________________________

Will a building permit be needed before the temporary power can be hooked up? □ YES □ NO

Is commercial electricity required in all homes in your area? □ YES □ NO

Can photovoltaic panels be used to supply your electrical needs? □ YES □ NO

Is there a phone accessible close to the site? □ YES □ NO

Will a site phone be available?[3] (It is highly recommended.) □ YES □ NO

What are the regulations for installing a temporary pole?

____

____

____

When can the phone be installed? ______________________________

Who will be responsible for clearing trees to make room for power lines? ______________

____

____

____

Notes

1. The cost of bringing electrical power to the land can be prohibitive. Some land may be selling for a greatly reduced price because the seller knows this. If there are no power lines near the site and there is not a high growth and development rate in your immediate vicinity, the power company may require you to pay the *entire* cost of bringing power to your property. This can run from $1,000 to $30,000. Also, they may require you to do all the clearing through the forest for any poles that may be needed. You may be able to contest this as they are responsible for supplying citizens with power, but check it out first.

2. Utility companies are notorious for not keeping agreements. Two weeks in their timetable often can look like two months to others. If you are running on a close timetable, and need to have power on the site for the work to progress quickly, be sure to give the phone and utility companies at least two to three months notice. Then keep in contact with them to be sure they are doing what they promised. I have all too often seen entire housing starts postponed because the power company did not deliver temporary power in time.

3. A site phone is a must, especially with an inexperienced builder. It is a means of immediate contact with the outside world: to gather information, coordinate deliveries and work schedules, order materials, and do all the thousands of things that take immediate phone communications.

Additional Information

Additional Information

5/Driveways, Roads and Bridges

A bulldozer driver that is sensitive to the land is a rare and precious find.

GOOD ACCESS to your home is vital. Living at the end of a muddy, rutted road not only shortens the life of the car, but may run your repair bills, both for the road and the car, into thousands of dollars. Aside from that, building a good road or bridge can also be expensive and may make the project unfeasible by diverting much of the funds that could have been spent directly on the house. We have sometimes advised people not to buy a certain piece of property or chose a certain site because of the cost of providing good access.

Your access is not only vital for convenience sake, but should contribute to the aesthetics and appeal of the approach to your home. An access that creates both a sense of privacy and of welcome is desirable. Though a straight road is the least expensive way to go, a gentle curve will only cost a small amount more and will serve to block noise and headlights and afford the home more privacy. Give a lot of consideration to the design of your road or driveway, as it can have a large impact on your home.

Driveways, Roads and Bridges

Will any roads or driveways need to be built to the site?[1] ☐ YES ☐ NO

How long will the road or driveway be? ____________________

What are the % slope(s) of any grades the road must climb?[2]

Grade #1 ____________

Grade #2 ____________

Grade #3 ____________

Grade #4 ____________

What are the local requirements for roads and driveways?

____

____

____

Will the road need to be asphalted or concreted?[3] ☐ YES ☐ NO

What is the maximum grade the city, county, or fire department will allow?

____

What are the road widths required by local government agency?

____

____

Is there a problem where the road joins the county or city road? ☐ YES ☐ NO

If yes, explain

____

____

Will a turn around be necessary for fire trucks? ☐ YES ☐ NO

Is the soil under the proposed road:

☐ Good
☐ Fair
☐ Poor

Special Conditions ____________________

How many drain culverts will be needed, and where?

Culvert #1 ______

Culvert #2 ______

Culvert #3 ______

Culvert #4 ______

Who installs and maintains culverts at main roads? ______

At bridges? ______

Will the road bid include culverts? ☐ YES ☐ NO

How many large trees will need to be cut? ______

Can the trees that are cut to make room for the road be used for lumber or firewood? ☐ YES ☐ NO

Are there any problem spots (low places, bad curves, bad drainage)?

Problem #1 ______

Problem #2 ______

Problem #3 ______

Will soft spots and low areas be built up? ☐ YES ☐ NO

Will any dirt need to be removed? ☐ YES ☐ NO

If yes, where will it be dumped? ______

Will any dirt be needed for fill? ☐ YES ☐ NO

If yes, where will it be taken from? ______

Will fill, cut, and dump areas need to be inspected prior to beginning work? ☐ YES ☐ NO

Will a permit need to be obtained? ☐ YES ☐ NO

If yes, what are the regulations of the permit?

______

______

______

What type of road is planned?

- □ Dirt
- □ Gravel
- □ Asphalt
- □ Concrete
- □ Other

How thick will the base rock layer be?[4] ______________________________

Is this the proper rock for the type of road you are planning? □ YES □ NO

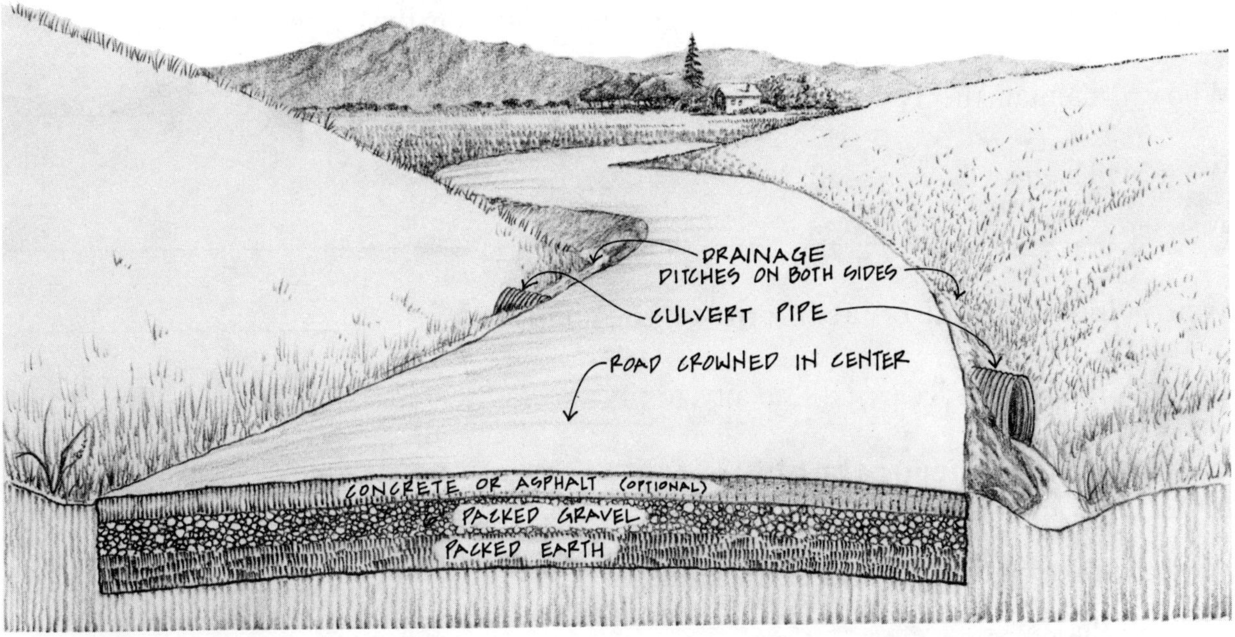

Will a vibrator-compactor be used to adequately compact the road?[5] □ YES □ NO

What is the cost per ton of road rock (including hauling to your site)?[6] ______________

Does the road need to be oiled?[7] □ YES □ NO

Is road oiling allowed by environmental agencies? □ YES □ NO

Are retaining walls needed? □ YES □ NO

What can be used for erosion protection?

Are the backhoe and dozer drivers sensitive to your land and needs? □ YES □ NO

Can on-the-site materials be used? □ YES □ NO

Who will perform and pay for compaction tests if any are needed? ____________________

If underground power, gas, phone, or sewer is to be put in, will it disturb the newly constructed road?[8] □ YES □ NO

Can concrete pavement blocks that allow grass to grow in parking areas be used? □ YES □ NO

Will the road lend privacy and keep headlights and noise from the house? □ YES □ NO

Who will maintain the road? ____________________

What road signs will be needed?

____

Who is legally liable for accidents on the road? ____________________

Can your road be deeded to state, county, or city? □ YES □ NO

Will any bridges need to be built? □ YES □ NO
If yes, what are the contractors' quotes?[9]

Contractor #1: ____________________
Phone ____________________
Quote ____________________
Type of bridge ____________________

Contractor #2: ____________________
Phone ____________________
Quote ____________________
Type of Bridge ____________________

Will engineering be needed for the bridges? □ YES □ NO

Can culverts take the place of bridges?[10] □ YES □ NO

What size and type of rock will be used for under-bedding?[11]

Size ______________________________

Type ______________________________

Contractor information

Contractor #1: ______________________________

Name ______________________________

Address ______________________________

Phone ______________________________

Quote ______________________________

Depth of base rock planned ______________________________

Type of road ______________________________

References ______________________________

Contractor #2: ______________________________

Name ______________________________

Address ______________________________

Phone ______________________________

Quote ______________________________

Depth of base rock planned ______________________________

Type of road ______________________________

References ______________________________

Notes

1. Do not underestimate the tremendous costs of roads or bridges, especially if much clearing and grading is needed, or if you will need to apply either asphalt or concrete. It may cost many thousands of dollars. Know about these costs before you buy the property and be sure you can afford the added expense.

2. The slope of the land can have a definite influence on the way that the road must be built, its cost, and how well it will perform in the future. Local agencies may require that the road or driveway be asphalt or concrete if the slope exceeds 15%. This is to assure that fire trucks can have access to your house.

Even if slope is not regulated, it may be wise to consider asphalt and concrete, as roads that are too steep can be troublesome in rainy and icy weather. This was the case with a house I built some years ago in La Honda, California. The road contractor built the driveway at a steep angle, instead of traversing the hill as we had requested. Even during the dry season, concrete trucks spilled part of their loads because of its steepness and supply trucks could not make the grade unless they had a heavy load to provide traction. Finally, the road was asphalted at an additional cost of $3,300.00 and even then it was difficult to maneuver in the rainy season when wet leaves coated the road. This road had other problems, such as curves that were improperly banked and loose base rock. We later found out that the road contractor had been building runways at the Oakland airport and knew little about building mountain roads where, unlike runways, curves and hills were a problem. The road, which was almost one mile long, eventually cost the owners over $22,000.00 and its quality was still in doubt.

3. Most roads are either dirt, gravel, asphalt, or concrete. Unless you live in a very dry area with stable soil, dirt roads are temporary at best, and soon ruts and potholes will appear. Gravel roads are popular in rural areas and are the least expensive way to go with a permanent road. Though they do not have the comfort and lack of maintenance of asphalt and concrete drives, they are much less costly and if well built can serve well with occasional maintenance. As mentioned earlier in this chapter, if the road or drive is too steep, asphalt or concrete is recommended and sometimes required.

Asphalt roads and drives, though cheaper than concrete, can often serve as well. Usually they are trouble-free during the first years of their life but may start to have minor (sometimes major) problems after the first five to ten years. They are advised in areas that have poor soil, or heavy rains or snows, or where there are steep slopes and you want to forego the cost of a concrete drive. The concrete drive is the best as well as the most expensive and is often not needed. It is nearly problem-free if the soil underneath is stable and well compacted and the road itself is well built with the proper slope and drainage.

4. Before accepting any quote on the job, be sure that you know how many inches of base rock are to be applied. Usually six inches or more is a good application. Some builders may reduce this to three to four inches in order to give a low quote.

5. When building roads, especially gravel roads, it is best to compact the soil as well as the gravel as thoroughly as possible to help hold the road together. Compacting is done most effectively by an expensive piece of equipment called a vibrator roller—one of those machines with a huge drum on the front that would roll over cartoon characters and smash them flat. Due to its cost many road builders never buy one and because of this the quality of their roads may suffer. Find out if one is recommended in your area and if your road builder has access to one.

6. With gravel roads a large part of the cost will be for the gravel itself and the hauling of the gravel. Therefore your distance from the gravel quarry has an impact on the cost of the road. Try to locate the closest quarry that has high quality rock.

7. Oiling a road is a process by which old oil is periodically poured over a gravel surface road to help it bond together and resist rain. It is sometimes needed in rainy areas where steep grades are involved. Oiling can have a high impact on the local environment, it can be costly, and the road may need to be re-oiled every few years. You may decide to go to an asphalt road to avoid this. Talk to your road builder and your neighbors before deciding. Don't make a quick decision; the ramifications can have a great effect on your pocketbook as well as on your car.

8. Be careful in dealing with the utility companies, whether they are bringing their lines in overhead or below ground. They are infamous for destroying land, trees, and roads without a second thought. A student of ours in North Carolina came home one day to find that the utility had just installed an underground wire along the left hand tire track of the road, rendering the road useless. Their reason was it was the simplest area for their ditch digger. Had the student not fought his way up to see a vice president with the company, the costs of repairing the road would have been his alone. Whenever utility company workers are on your land, be sure you are present.

9. Bridging over what appears to be a small creek can be surprisingly costly. Remember that the bridge must be designed and built to withstand any possible flooding, as well as carry the weight of concrete and supply trucks during construction. Do not be fooled into thinking that the bridge crossing that peaceful little brook in your front yard will be cheap.

10. If you are lucky you may be able to use a large culvert pipe instead of building a bridge, especially if the creek is small. Check out this possibility—it can save you much money.

11. Different soil profiles sometimes influence the size and makeup of the base rock to be used. Also the local climate, especially rainfall, may influence this. In each area it differs; be sure that whoever is building the road understands the different factors that go into deciding which base rock should be used and chooses the proper one. I recently saw tons of rock spread on a road only to find out later that it was of a nature that would easily break into dust under much weight.

Additional Information

6/Waste Systems

IF YOU ARE hooking into an existing sewer system, the unknowns are few and the operation relatively simple (note that I said relatively). Providing your own waste system in rural and many suburban areas can be a good bit more difficult, and in areas where the soil is unsuitable for septic systems, can threaten the feasibility of the entire project. In either case there are certain things you need to know about the disposal of your wastes.

Both septic systems and central waste systems have problems, economically and ecologically. They are costly and require maintenance which you pay for, either directly or through your tax dollar. Both kinds of systems pollute the local environment and waste a tremendous amount of water in disposing of a small amount of waste (approximately 25,000 gallons of clean water is needed to dispose of 166 gallons of human waste.) More efficient and environmentally safer systems have been developed and are gaining acceptance and legality in the United States. You may want to consider them as possibilities.

This chapter has been divided into two sections: hooking into existing systems, and developing your own septic systems. Some ecologically sound alternative waste systems are also mentioned towards the end of the chapter.

For Further Reading

Goodbye to the Flush Toilet. Carol Hupping Stoner. Emmaus, PA: Rodale Press, 1977.

Residential Water Re-Use. Murray Milne. California Water Resources Center, University of California, Davis, CA 95616.

Septic Tank Practices. Peter Warshall. New York: Doubleday & Co., 1979.

Existing Sewer Systems

If the property is in a subdivision, do sewer outlets exist?[1] □ YES □ NO

What size pipe is used? ____________________

Who is allowed to tie in to the existing system?[2] ____________________

How deep is the tie-in pipe? ____________________

What type of pipe is allowed? ____________________

What parts are needed to connect to the existing system? ____________________

____

Who will pay for these? ____________________

Is a permit needed for a connection? ____________________

If you are tapping in to an existing sewer system, what are the tap-in fees? ____________________

The monthly fees? ____________________

Developing Your Own Septic System

Will a septic system need to be built? □ YES □ NO

If not, what are the costs of hooking into the existing systems? ____________________

Are there yearly fees? $____________________

Are there any existing sewer moratoriums in the area? □ YES □ NO

Are there any pending moratoriums? □ YES □ NO

How well does the soil perc?

- □ Good
- □ Fair
- □ Poor

Can you do your own perc test? □ YES □ NO

What type of septic systems are used in your area? ____________________

____

Are there any problems with septic systems in the area? ☐ YES ☐ NO
If yes, what are they?[3] ______________________________

____

____

What is the life of the average septic system in your area? ______________________________

Are systems in your area subject to clogging? ☐ YES ☐ NO

If yes, should two leach beds be considered? ☐ YES ☐ NO

How large a system will be required for your home?[4] (See Table 2, page 86.)

Tank ____________________

Leach lines ____________________

Will the septic system become overloaded if you increase the size of the house and thereby increase the demands on the system? ☐ YES ☐ NO

How far will your septic tank be required to be from the well? ______________

From the house? ____________________ (See Table 1, page 85.)

Are there any nearby water sources (creeks, streams, lakes, etc.) that could prohibit the use or restrict the location of the septic area? ☐ YES ☐ NO

Can distances from the well to the septic system be reduced by using additional casing or grouting in drilled wells? ☐ YES ☐ NO

Are there any steep grades near the septic tank site that could cause seepage problems?[5] ☐ YES ☐ NO

If so, where are they? ______________________________

____

____

Is there any impervious rock below soil level that could cause a problem? ☐ YES ☐ NO

What type of septic tank do you plan to use? ______________________________

Can you build your own tank? ☐ YES ☐ NO

Where will the leach lines run?[6]

____

____

____

Will sewer lines or septic tanks cut heavily into any tree root systems? □ YES □ NO

Are the sides of the leach field trenches being glazed during digging?[7]
□ YES □ NO

Is the planned septic area in a flood plain or does it have other drainage problems?
□ YES □ NO

Will any part of the septic field be under the house? □ YES □ NO
The drive or road? □ YES □ NO
Other structures? □ YES □ NO

Are there any large trees that must be cut to make space for the tank or field? □ YES □ NO

If yes, how many? ____________________

Will an engineered system be needed? □ YES □ NO

Will a standard system work or will a vacuum, pressure, or pumping system be required? □ YES □ NO

Will the system be uphill or downhill from bathrooms? (Uphill can cause problems.) ____________________

Can the sewer share a ditch with the:[8]

- □ Power
- □ Phone
- □ Gas
- □ Water
- □ Cable television

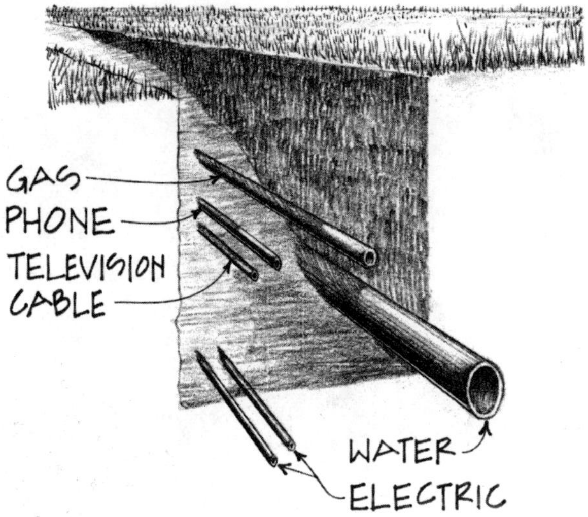

What is the required slope of the sewer line?[10] ____________________

Do any of the following potential problems exist?[10]

- □ High water table
- □ Too close to well
- □ Too close to steep grade
- □ Too close to property line
- □ Compacted clay or rocky soil
- □ Inaccessible to large equipment
- □ Shortage of contractors

Will the subsoil and topsoil be separated if a septic system or trenches are to be put in?[11] □ YES □ NO

What is the cost of having the septic tank pumped? ____________________

How often will pumping be necessary? ____________________

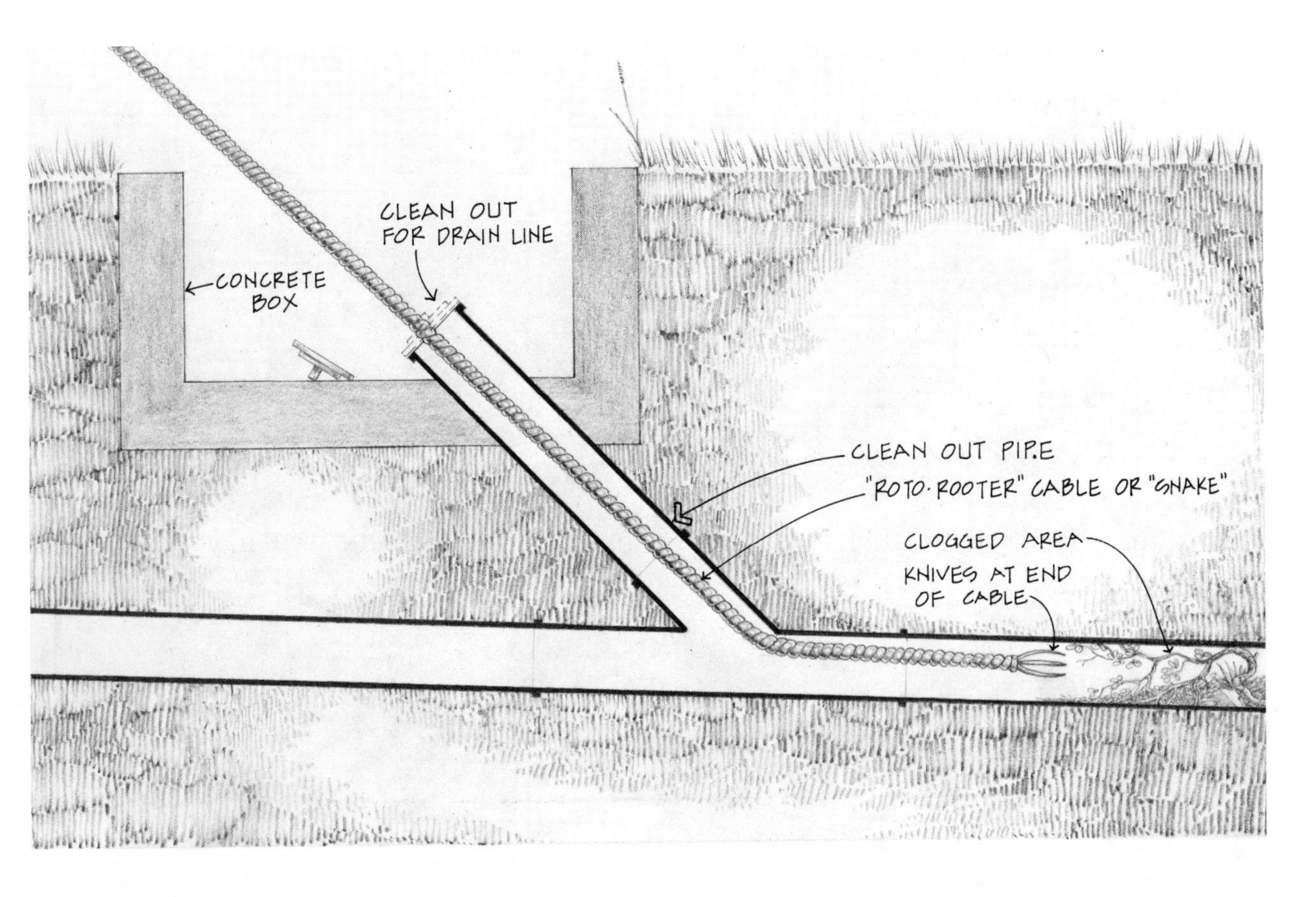

What clean-outs are required?[9] ______________________________

Will any chemicals or objects be dumped into the sewer system that could effect the biological degrading action? □ YES □ NO

Where will the septic sludge or final waste product be dumped? ______________

Septic Contractors

Septic contractor #1:

Name ______________________________

Address ______________________________

Phone ______________

Quote ______________

Type of tank ______________

Size of tank ______________

Length of leach line ______________

Septic contractor #2:

Name ______________________________

Address ______________________________

Phone ______________

Quote ______________

Type of tank ______________

Size of tank ______________

Length of leach line ______________

Are alternative systems, such as composting or incinerating toilets, legal in your area?[13]

□ YES □ NO

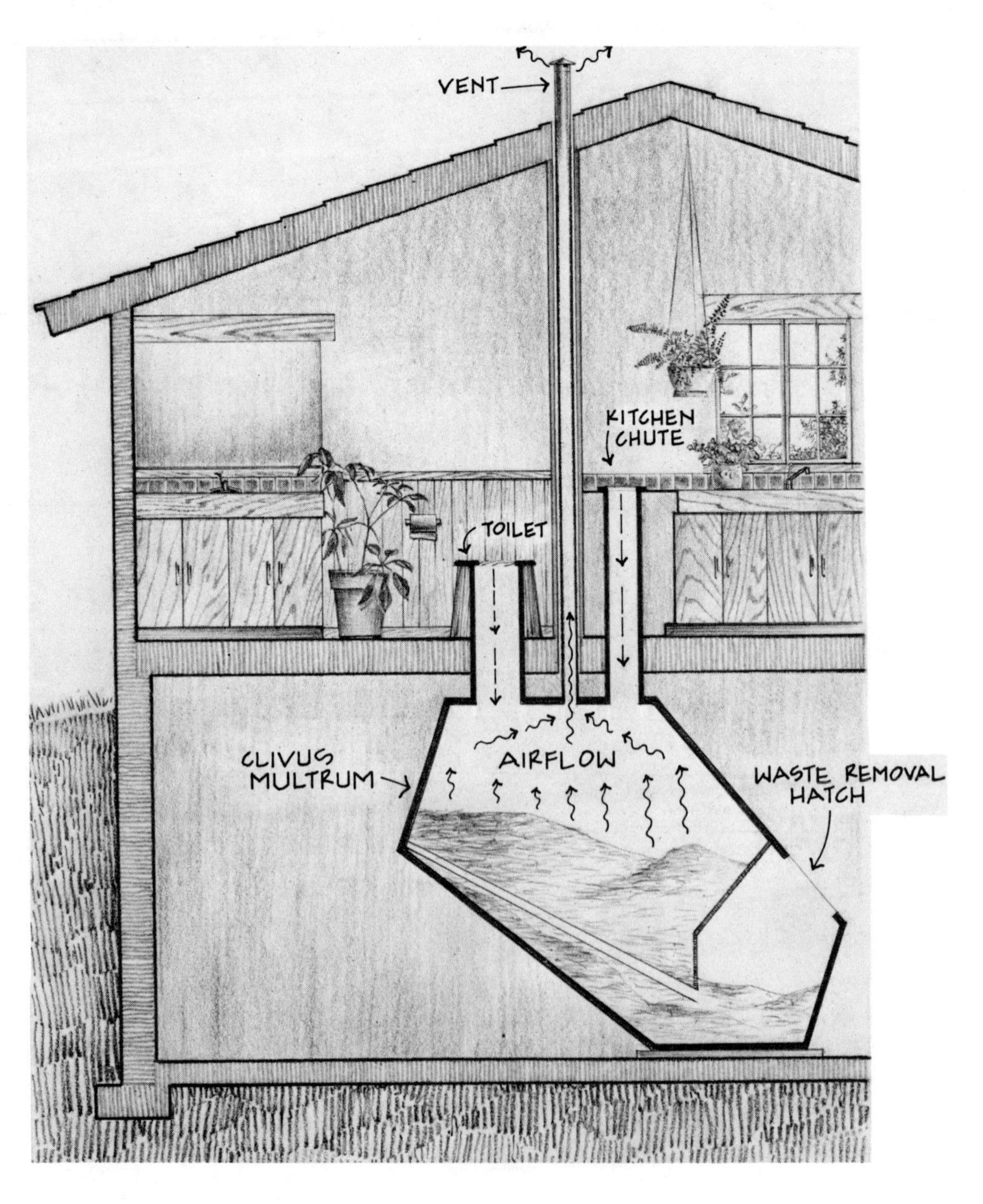

Notes

1. Subdivision lots are often sold with water and sewer hookups supposedly already in place. However, good maps as to the exact locations, or sometimes the pipes themselves, are nonexistent. Responsibility for any additional costs incurred by these missing or lost hookups is often in doubt. We recently built a house in a subdivision put together by a developer. Many of the hookups were never in place and one was put in backwards. The developer had long gone and we had to cover the costs.

2. It is important to know who may tie in because in a central system people are tied together and one person's mistake may have consequences for all. It is often required that a licensed plumber do the hookup from the home to the main sewer drain. Sometimes owners or unlicensed people are allowed to do their own work—check the regulations in your area.

3. To best find out about septic tank problems, ask your neighbors, the local septic tank contractors, and your local health board. Usually most problems relate to the porous qualities of the soil, or the local water table.

4. The size of the tank and the length of the leach lines are usually spelled out by the local building or health department. They are calculated by how well the soil percs, and in accordance with how many bathrooms or bedrooms you plan to have. Calling a spare bedroom a studio may sometimes reduce the requirements of the septic system. You may want to find this out before you label your plans. The *One and Two Family Dwelling Code* requires a 750 gallon tank for a 1 or 2 bedroom house; a 1000 gallon tank for a 3 bedroom; and a 1200 gallon tank for a 4 bedroom house.

5. It is usually best not to build any part of the septic system in any area that has a steep grade. The effluent water will run towards the lower end of the leach lines and they will become waterlogged while the upper ends will not be used at all. Also, it is wise (and sometimes regulated) not to build within 75–150 feet of a steep grade because the leaching effluent may hit an impervious layer of rock or soil and be transported horizontally underground until it oozes out the side of the slope.

6. Do not randomly place your leach lines. First consider the following: existing structures; future structures; gardens; roads and drives; major trees and their root systems; location from house and well. Place the tank and lines according to other features of the property and decide where you may want a swath cut on your land. It can later be replanted but there may be an area you are planning to clear in the future and proper placing of the septic tank and leach lines may prevent added clearing work later.

7. If the soil is very moist, the sides of the leach field trenches may become glazed or smoothed over as the backhoe bucket digs the trench. This greatly inhibits the soil from allowing the effluent water to pass through and actually acts as a bowl which will hold water. Landscapers know trees will die if placed in a hole with glazed sides because they will drown if the hole is holding water. Be sure the sides of the trenches are roughed up with a rake after they are dug to assure good leaching. It was discovered recently that more leaching happens through the sides of the trench than the bottom because the bottom soon becomes impacted with solids from the effluent water. In some states, backhoe operators are required to weld knobs on the sides of their buckets to insure that they do not glaze the sides of the leach field trenches.

8. Rather than digging separate ditches for all the utilities, water, etc, find out if it is possible for them to share a ditch. Very often this is allowed with some restrictions on how deep the ditch has to be, how far apart the different pipes or wires must be from each other, and how far below the surface they must be. Also you may want to consider putting a coolth tube in the ditch to help cool the building.

9. Clean-outs are plugs leading into the sewer lines that allow you to get a cleaning tool called a "snake" into the line, should it get clogged. There are often requirements about how frequently clean-outs must occur in any run of sewer pipe. They usually protrude above ground, or are set in concrete boxes below the surface.

10. In order for the effluent waste and water to be carried from the house to the septic system or sewer, the drain pipe must have a certain slope to it. This is usually regulated by the health or building department. If the house is a great distance away from the sewer system, the ditch for the pipe may need to be very deep and thereby very costly. Usually the slope is 2–4″ drop for every 100 feet.

11. Subsoil, approximately 6–12″ below the surface, is sterile and nothing will grow in it. Only the topsoil, consisting of composted materials of organic matter, (i.e. tree leaves, limbs, etc,) can support plant growth. Topsoil that is being dug up should be separated and used again during the landscaping stage. Subsoil should also be separated and care taken that it does not get spread over the surface. The reason new house sites often look sterile and the lawns seem to be having such a hard time getting started is that subsoil was spread on the land after construction was completed and before landscaping was begun.

12. "You may want to research if there will be any future legislation that may effect where your waste is deposited. The Clean Water Act, for instance, plans to eliminate all pollutant discharge into waterways by 1985." From *Toilet Papers* by Sim Van de Ryn.

13. For both economic and ecological reasons you may want to consider what is known as a waterless toilet. The best brand is a Clivus Multrum which has been used in Sweden successfully for years and is now becoming more popular in the United States. Both Maine and counties of California have approved its use as the only toilet system in the house. It utilizes a large fiberglass container located beneath the house, or on the first floor of a two story house, that receives both toilet and kitchen wastes. Gray water is treated in a separate system.

The toilet has no flush mechanism and thereby saves 30% of the household water use. Waste products are decomposed by bacteria and used as compost.

Additional Information

Table 1. Absorption Field Design

State	Setbacks		Minimum Spacing (Feet)	Minimum Cover (Inches)	Minimum Percolation Restrictions	Trench Widths (Inches)	Sizing
	Well	Surface Water					
Alabama	50-75		6	6	None	18-36	Perc.
Alaska	50-100	50-100	6	12	None	12-36	Perc. and Soils
Arizona	50-100	100	6	12	None	12-18	Perc.
Arkansas	–	–	–	–	–	–	–
California	–	–	–	–	–	–	–
Colorado	100	50	6	12	None	18-36	Perc.
Connecticut	75	50	6-9	6	None	18-36	Perc.
Delaware	–	–	–	–	–	–	–
Florida	75-100	50	6-8	12	None	18-24	Perc. and Soils
Georgia	100	50	10	12	None	18-36	Perc. and Soils
Hawaii	–	–	–	–	–	–	–
Idaho	100	100-300	6	12	None	12-36	Perc. and Soils
Illinois	–	–	–	–	–	–	–
Indiana	50-100	50	6-7.5	12	None	18-36	Perc.
Iowa	100-200	25	7.5	12	None	18	Perc.
Kansas	–	–	–	–	–	–	–
Kentucky				None			Perc.
Louisiana	100			6-12	None	12-18	Perc.
Maine	100-300	50-100	10	2-6	None	24	Soils
Maryland	–	–	–	–	–	–	–
Massachusetts	–	–	–	–	–	–	–
Michigan	–	–	–	–	–	–	–
Minnesota	–	–	–	–	–	–	–
Mississippi	–	–	–	–	–	–	–
Missouri	–	–	–	–	–	–	–
Montana	100	100	6	12	Yes	12-36	Perc. and Soils
Nebraska	100	50	6	6	No	18-36	Perc.
Nevada	100	100	6	4-6	Yes	12-24	Perc.
New Hampshire	75	75	6-7.5	6	None	12-36	Perc.
New Jersey	–	–	–	–	–	–	–
New Mexico	–	–	–	–	–	–	–
New York	–	–	–	–	–	–	–
North Carolina	–	–	–	–	–	–	–
North Dakota	–	–	–	–	–	–	–
Ohio	50		6	6	None	8-30	Soils
Oklahoma	–	–	–	–	–	–	–
Oregon	50-100	50-100	10	6	None	24	Soils
Pennsylvania	100	50	6	12	Yes	12-36	Perc.
Rhode Island	100	50	6	12	None	18	Perc.
South Carolina	–	–	–	–	–	–	–
South Dakota	100	100	6		Yes		Perc.
Tennessee	50	50	6	12	None	18-36	Perc. and Soils
Texas	–	–	–	–	–	–	–

Continued on page 86.

Table 1. Absorption Field Design (continued)

State	Setbacks Well	Setbacks Surface Water	Minimum Spacing (Feet)	Minimum Cover (Inches)	Minimum Percolation Restrictions	Trench Widths (Inches)	Sizing
Utah	100	100	6-7.5	12	None	12-36	Perc.
Vermont	–	–	–	–	–	–	–
Virginia	55-100	50-100	6-9	None	None	18-36	Perc. and Soils
Washington	75-100	100	6	6	Yes	18-36	Perc. and Soils
West Virginia	100	100	6	12	None	12-36	Perc.
Wisconsin	50-100	50	10	12	None	18-36	Perc. and Soils
Wyoming	100	50	6-7.5	6-12	None	12-36	Perc.

Table 2. Septic Tank Design and Water Depth

State	Tank Size in Gallons, Number of Bedrooms: 1	2	3	4	5	Minimum Water Depth (Feet)	Open Discharge
Alabama	1000	1000	1000	1200	1400	4	
Alaska	750	750	900	1000	1250	4	No
Arizona	960	960	960	1200	1500	4	No
Arkansas	–	–	–	–	–	–	–
California	–	–	–	–	–	–	–
Colorado	750	750	900	1000	1250		Yes
Connecticut	1000	1000	1000	1250	1500	1.5	No
Delaware	–	–	–	–	–	–	–
Florida	750	750	900	1000	1200	1.5	No
Georgia	750	750	900	1000	1250		No
Hawaii	–	–	–	–	–	–	–
Idaho	750	750	900	1000	1250	4	No
Illinois	–	–	–	–	–	–	–
Indiana	750	750	900	1100	1250	–	No
Iowa	750	750	1000	1250	1500	1.5[A.]	Yes
Kansas	–	–	–	–	–	–	–
Kentucky	750	750	900	1000	1250		No
Louisiana	500	750	900	1150	1400	None	Yes
Maine	750	750	900	1000	1250	2	Yes
Maryland	–	–	–	–	–	–	–
Massachusetts	–	–	–	–	–	–	–
Michigan	–	–	–	–	–	–	–
Minnesota	–	–	–	–	–	–	–
Mississippi	–	–	–	–	–	–	–
Missouri	–	–	–	–	–	–	–
Montana	750	750	900	1000	1250	4	No
Nebraska	750	750	900	1000	1250		No
Nevada	1000	1000	1000	1000	1250	4	No
New Hampshire	750	750	900	1000	1250	4[A.]	No
New Jersey	–	–	–	–	–	–	–

Continued on page 87.

Table 2. Septic Tank Design and Water Depth (continued)

State	Tank Size in Gallons Number of Bedrooms 1	2	3	4	5	Minimum Water Depth (Feet)	Open Discharge
New Mexico	–	–	–	–	–	–	–
New York	–	–	–	–	–	–	–
North Carolina	750	750	900	1000	1250	2[A.]	No
North Dakota	–	–	–	–	–	–	–
Ohio	1000	1000	1500	2000	2000	4[A.]	Yes
Oklahoma	–	–	–	–	–	–	–
Oregon	750	750	900	1000	1250	1.5[A.]	No
Pennsylvania	900	900	900	1000	1100	4	No
Rhode Island	750	750	900	1000	1250	3	No
South Carolina	–	–	–	–	–	–	–
South Dakota	1000	1000	1000	1250	1500	4	No
Tennessee	750	750	900	1000	1250	4[A.]	No
Texas	–	–	–	–	–	–	–
Utah	750	750	900	1000	1250	1	No
Vermont	–	–	–	–	–	–	–
Virginia	30 hour Detention 100 Gallons Per Day					No Minimum	Yes
Washington	750	750	900	1000	1250	3[A.]	No
West Virginia	750	750	900	1000	1250	4	No
Wisconsin	750	750	975	1200	1375	3[A.]	No
Wyoming	750	750	900	1000	1250	4	Yes

Table 3.

Soil Texture	Estimated Percolation Rate Range Inches/Minutes
Sand	0-10
Sandy Loams, Loams	10-30
Porous Silt Loams, Silty Loams, Loams, Clay	30-45
Compact Silt Loams, Silty Clay Loams	45-90

Table 4. Minimum Absorption Requirements for Private Residences in Virginia

Percolation Rate (Time required for water to fall 1 inch in minutes)	Required absorption area, per 100 gallons/day of water used (in square feet)
1 or less	52
2	63
3	74
4	85
5	93
10	122
15	140
30	184
45	222
60	244

Over 60 minute rate will require special design which meets the approval of the Regional Sanitarian and/or the Director of the Bureau.

In every case sufficient area shall be provided for at least 400 sq. ft., unless otherwise indicated by soil evaluation.

Absorption area for standard trenches is to be figured as trench-bottom area.

Drainfield design should be based on peak loading and not on long time averages.

Table 5. Rated Absorption Capacities of Five Typical Soils (Table 25-D, One and Two Family Dwelling Code)

Type of Soil	Required Sq. Ft. of Leaching Area/100 Gals of Septic Tank Capacity	Max. Absorption Capacity Gals/Sq. Ft. of Leaching Area
(1) Coarse Sand or Gravel	20	5
(2) Fine Sand	25	4
(3) Sandy Loam or Sandy Clay	40	2.5
(4) Clay with Considerable Sand or Gravel	60	1.66
(5) Clay with Small Amount of Sand or Gravel	90	1.11

Table 6. Allowed Locations of Sewage Disposal Systems (Table 25-A, One and Two Family Dwelling Code)

Minimum Horizontal Distance in Clear Required From:	Building Sewer	Septic Tank	Disposal Field	Seepage Pit or Cesspool
Buildings or Structures[1]	2′	5′	8′	8′
Property Line Adjoining Private Property	clear	5′	5′	8′
Water Supply Lines	50′[2]	50′	50′	100′
Streams	50′	50′	50′	100′
Large Trees	–	10′	–	10′
Seepage Pits and Cesspools[3]	–	5′	5′	12′
Disposal Fields	–	5′	4′[4]	5′
Domestic Water Lines	1′	5′	5′	5′
Distribution Box	–	–	5′	5′

1. Including porches and steps whether covered or uncovered, breezeways, roofed porte-cocheres, roofed patios, carports, covered walks, covered driveways and similar structures or appurtenances.

2. All nonmetallic drainage piping shall clear domestic water supply wells by at least 50 feet. This distance shall be reduced to not less than 25 feet when approved type metallic piping is installed. Where special hazards are involved, the distance shall be increased as may be directed by the building official or public health authority.

3. When disposal fields and/or seepage pits are installed in sloping ground, the minimum horizontal distance between any part of the leeching system and the ground surface shall be 15 feet.

7/Permits

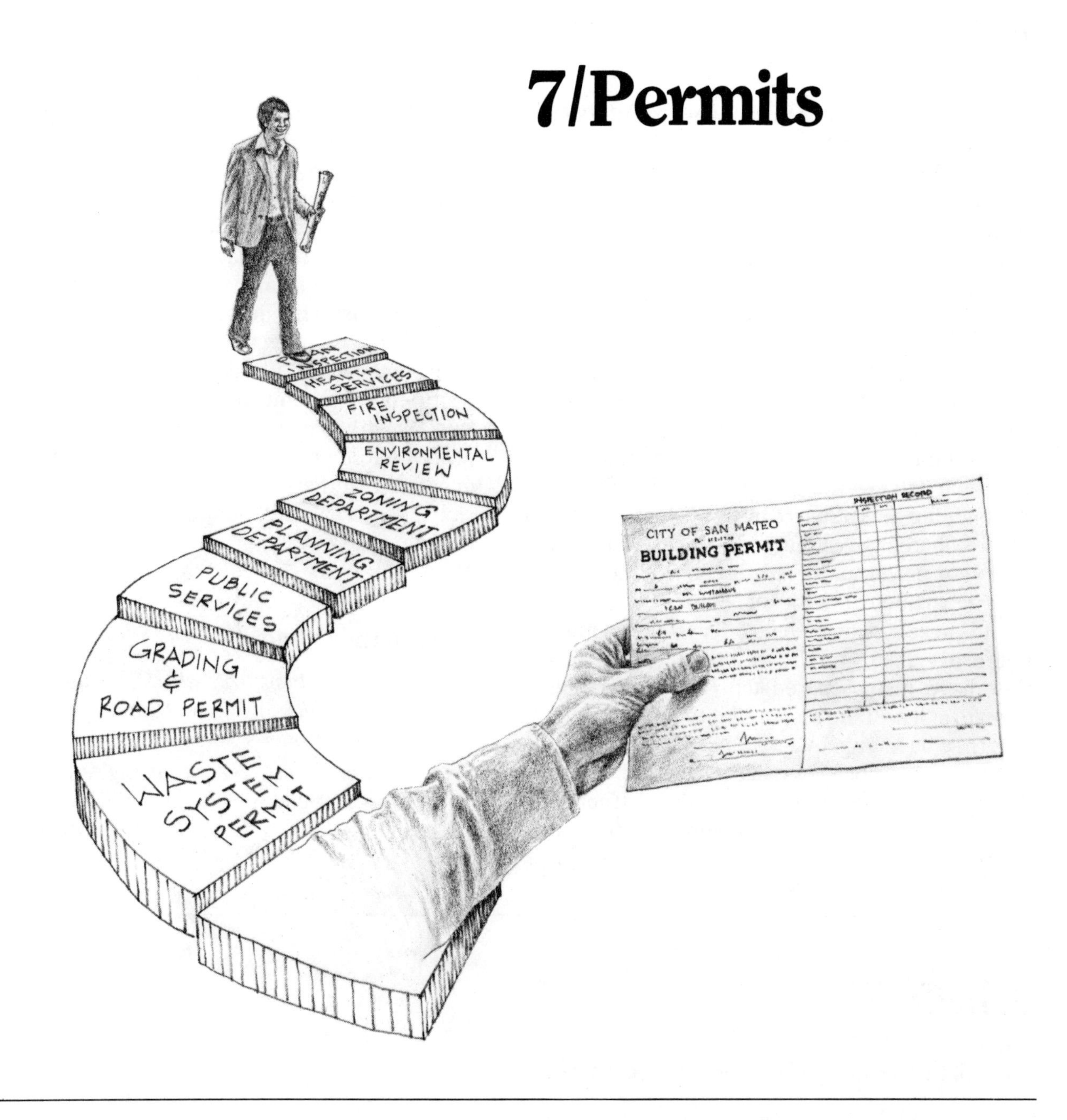

Understand the quickest and easiest way in which permits are obtained, and then use it.

GETTING YOUR building permit can involve anything from one or two quick visits to the building department to a six to twelve month emotional trauma that can leave you with the desire to take up arms against the government or abandon the entire project before building begins. Unfortunately, the ease or difficulty which you experience in getting your permit often depends more on your local regulatory agencies than the merits of your own project. It is wise, however, to be well prepared and know what to expect beforehand. The only advantage you have in dealing with the agencies is that you know you are coming and they don't. In some circumstances this allows you to do some investigative work to determine beforehand their likely responses to your requests. Knowing their responses may alter your approach or help you to prepare counter-responses. If you are building in an area where the permit process is known to be difficult, two things may help you maintain your sanity.

First, understand that most building departments are accustomed to dealing with professional builders and contractors who know what is needed to get a permit. Only occasionally (though it is becoming more frequent) do they encounter a novice. Usually there is little or no system set up to communicate to the novice what, exactly, is required in order to obtain a permit to build. The building departments often do not understand this need and leave you with a sense that they have just told you everything you need to know. And they believe they have, but you may find later that a few key things were not included. To avoid these difficulties, take personal responsibility for finding out what you will need to get a permit. Do not expect the building and other regulatory departments to accurately tell you. Find which departments must sign your permit paper, go directly to the head of each department, ask him or her for a written list of what is needed from you in order to have your permit form signed, and then ask them to sign the list they have just prepared for you. This requires them to at least take responsibility for the information they have given you.

Second, do not view your permit departments and their inspectors as "the enemy" (though it sometimes takes the compassion of a Zen monk not to do so). They are public servants, their job is to help you and if you view them and treat them this way they will remember this too. Be respectful and ask them when you need help. Try to include them in your project so that your success becomes their success and remember they are sometimes as frustrated by the bureaucracy as you are and they are not always to blame. At the same time, do not forget that assertiveness and even righteous anger can quicken the slow wheels of government. Your permit may sit in a basket for three weeks while waiting for five minutes of someone's time. If you can talk with that person it may speed up the entire process. Remember that government employees may feel that there is no quicker way you can possibly get your permit, but this does not mean there is none.

For Further Reading

Building Regulations: A Self-Help Guide for the Owner-Builder. Edmund Vitale. New York: Charles Scribner's Sons, 1979.

One and Two Family Dwelling Code. Council of American Building Officials, 560 Georgetown Building, 2233 Wisconsin Ave., Washington, D.C. 20007.

Permits

Checklist of possible permit requirements:

- ☐ Potable water on site ________________ GPM
- ☐ Road permit
- ☐ Grading permit
- ☐ Environmental Impact Report
- ☐ Perc test or septic permit
- ☐ Energy code calculations
- ☐ Coastal Commission permit
- ☐ Flood Control permit
- ☐ Fire department O.K.
- ☐ Encroachment permit
- ☐ Workman's compensation insurance application
- ☐ Fault zone O.K.
- ☐ Endangered species habitat
- ☐ Engineered foundation due to fault zone, slides, or poor soil

- □ Tree cutting permit
- □ Carport, garage, or offstreet parking permit
- □ Building height O.K.
- □ Zoning clearance
- □ Well drilling permit

Plans must include (check those required):

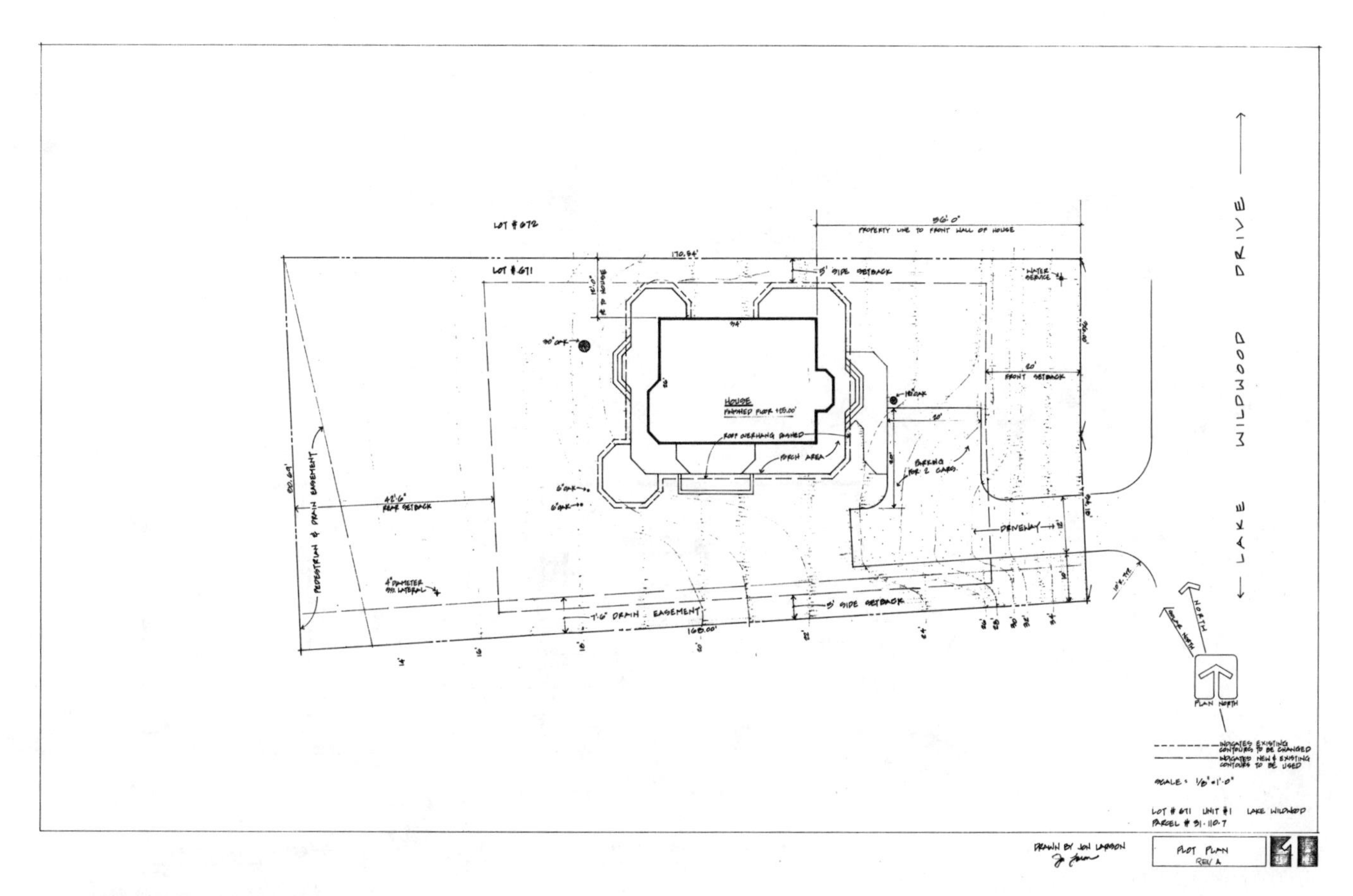

Plot Plan, including:

- □ Driveway
- □ Sewers
- □ Septic tanks
- □ % of land covered by building
- □ North arrow
- □ Scale of drawings
- □ Owner's name
- □ Street address
- □ Complete survey
- □ Tax assessor's parcel number
- □ Locations of buildings
- □ Dimensions of buildings
- □ Property line setbacks

- □ Distance between buildings
- □ Offstreet parking
- □ Existing buildings and their use
- □ Adjacent streets
- □ Slopes and distances of driveways
- □ Location of well or water line
- □ Other: ______________________________

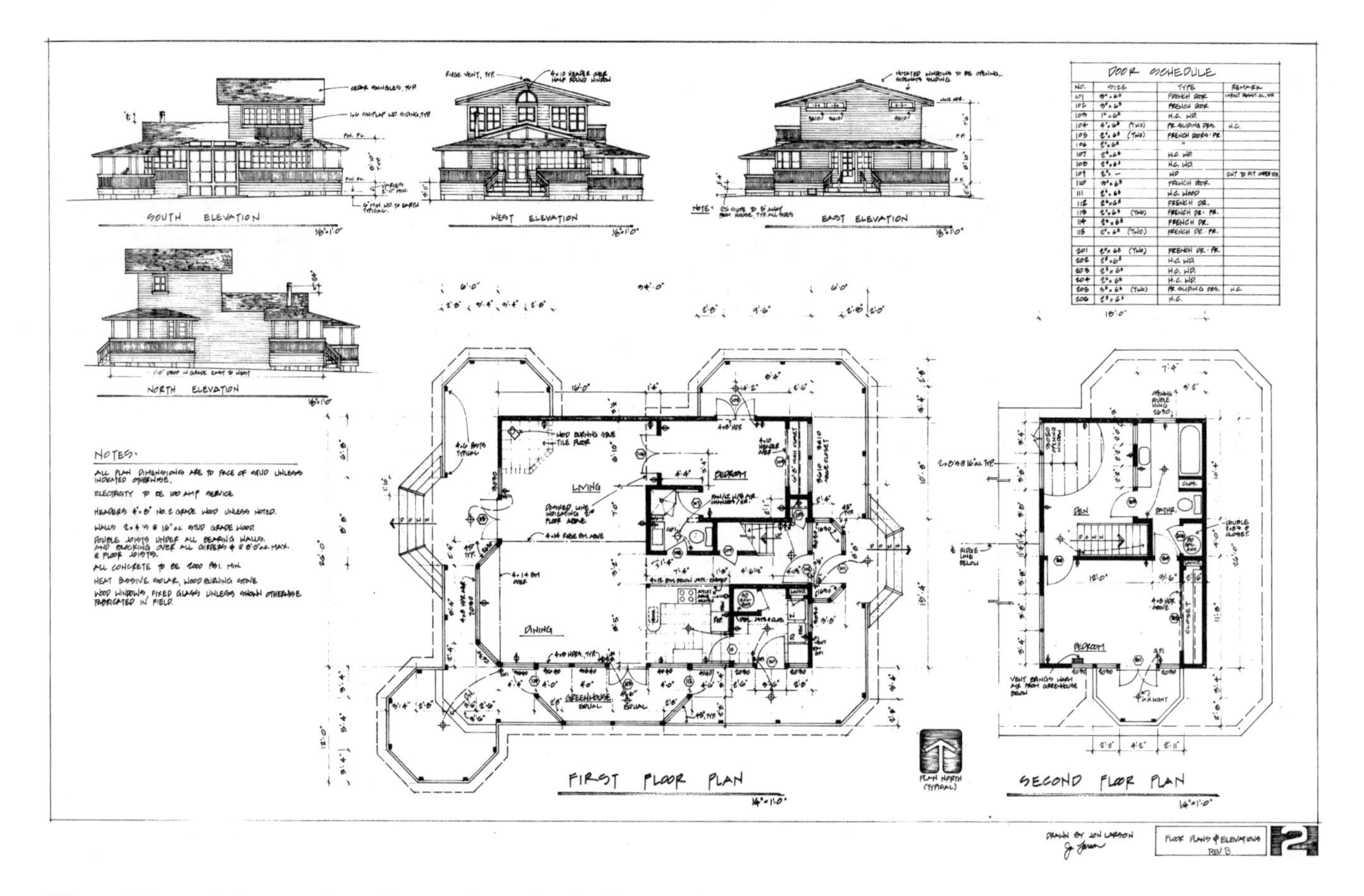

Floor Plan with complete dimensions, including:

- □ Early warning devices
- □ Appliances
- □ Heating systems
- □ Electrical outlets
- □ Fixtures
- □ Attic access
- □ Decks & porches
- □ Doors & windows, types, sizes, direction of swing
- □ Stairs & handrails

Main Structural Members, including:

- □ Direction & size of joists
- □ Direction & size of rafters
- □ Location & size of posts & girders
- □ Total floor area
- □ All information to show compliance with code

Exterior Elevations:

- □ Side elevations, how many? ____________________
- □ Height requirements

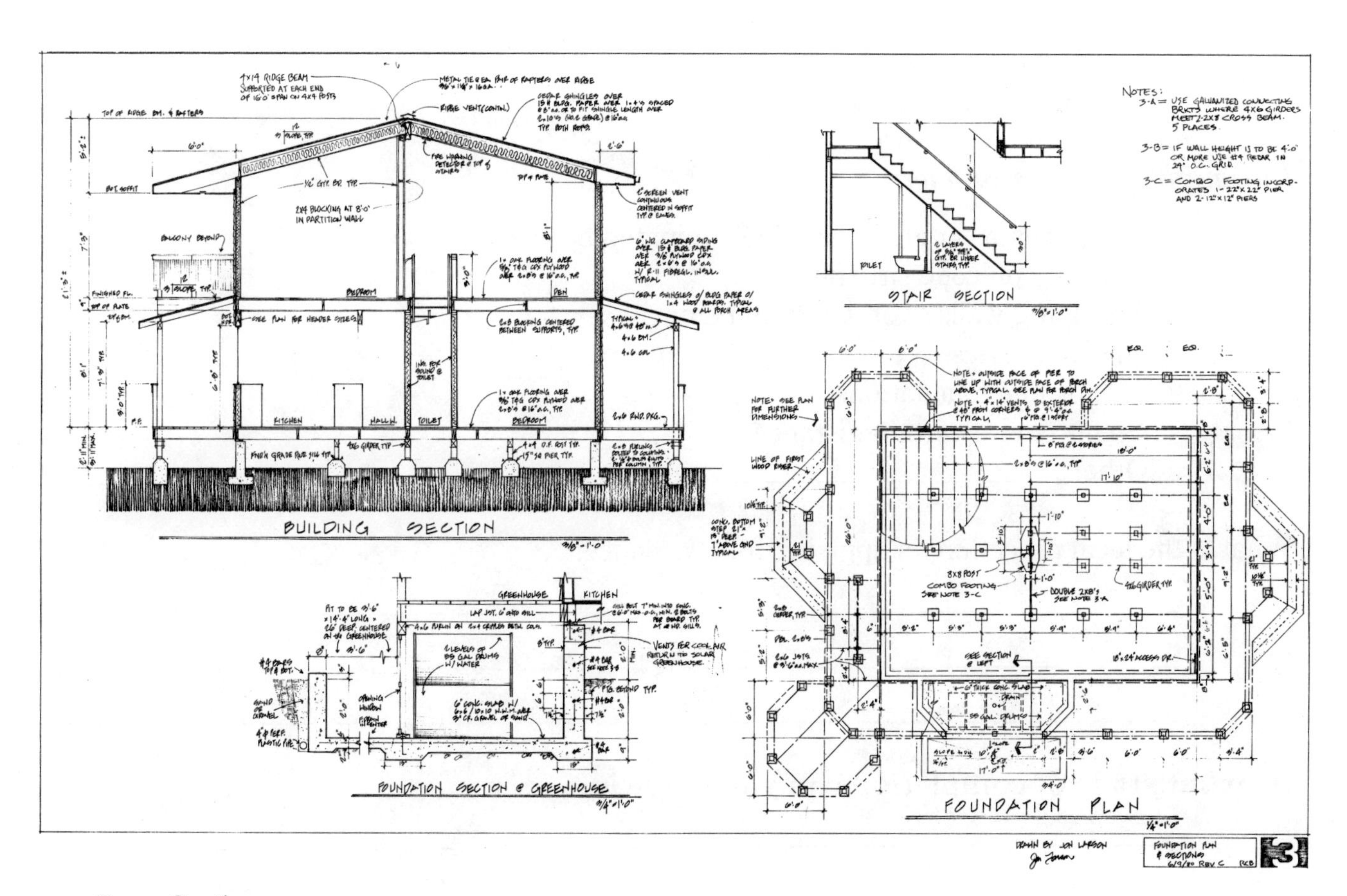

Cross Section:

- □ Structural layout
- □ Insulation placement
- □ Unusual details
- □ Framing plan

Insulation Plan:

- ☐ Weatherstripping
- ☐ Insulation in wall, roof, floor, ceiling
- ☐ Types used
- ☐ Supply & return ductwork
- ☐ Glass requirements
- ☐ Other: ______________________________

Miscellaneous:

- ☐ Road or driveway permit
- ☐ Bridge permit
- ☐ Copy of deed
- ☐ Water rights
- ☐ Solar shade rights
- ☐ Mining rights

Engineering may be needed for:

- ☐ Large beams & glu lams (glue laminated beams)
- ☐ Special non-code situations
- ☐ Retaining walls over a certain height
- ☐ Prefab trusses
- ☐ Pile or pile foundations
- ☐ Spans over a certain distance
- ☐ Other: ______________________________

What is the local procedure in applying for a variance?[1]

____

____

____

At what stage of completion can you move in?[2]

____

____

At what stage can you get your final inspection?

____

____

Will the bank give you a mortgage without a final inspection? ☐ YES ☐ NO

Will any existing structures need to be brought up to code? ☐ YES ☐ NO

If so, what will be required? ______________________________

____

____

Will you be taxed at a higher rate while your are building?[3] ☐ YES ☐ NO

Will the valuation for your tax base be adjusted because you are using your own labor? ☐ YES ☐ NO

Are any safety rules enforced on local jobsites? ☐ YES ☐ NO

____

____

How soon after you get your permit must you begin work?[4] ______________

Which of the following regulatory departments must sign your permit?

☐ Plan checker
☐ Engineering
☐ Fire
☐ Zoning
☐ Flood Control
☐ Coastal Commission
☐ Environmental
☐ Public Health
☐ Other ______________________________

What is the penalty for not obtaining a permit?[5]

Minimum: ______________________________
Maximum: ______________________________

Are any of the following permits required?

☐ Demolition of existing building
☐ Drilling for well
☐ Bridge
☐ Temporary power
☐ Fire or forestry
☐ Blasting
☐ Special use
☐ Road or driveway

What are total fees for all permits? ______________________________

How long are the permits good for? ______________________________

Are extensions for permits easily available? □ YES □ NO

What is the penalty for not finishing before the permits expire?

____

____

How long is the usual permit process? ______________________________

Will the permit process, the codes, or the tax structure alter in your favor if you change the names or designated uses of the buildings or any of the rooms?[6] □ YES □ NO

____

____

____

Will there be any union problems?[7] □ YES □ NO

Is an EIS (Environmental Impact Statement) required? □ YES □ NO

Are drainage or landscape plans required? □ YES □ NO

Are topographical maps required? □ YES □ NO

Will design require engineering? □ YES □ NO

Will an archeological permit be needed? □ YES □ NO

Departments involved in obtaining a permit:

BUILDING DEPARTMENT

Person contacted ____________________

Persons needed to sign permit ____________________

Location ____________________

Phone ____________________

Time factor till completion ____________________

Requirements:

1. ____________________
2. ____________________
3. ____________________

Other possible requirements:

1. ____________________
2. ____________________
3. ____________________
4. ____________________

Possible problems:

1. ____________________
2. ____________________
3. ____________________

Permit fees ____________________

Length of Permit ____________________

Departments involved in obtaining a permit:

ZONING DEPARTMENT

Person contacted ________________________________

Person needed to sign permit ________________________________

Location ________________________________

Phone ____________________

Time factor till completion ________________________________

Requirements:

1. ________________________________
2. ________________________________
3. ________________________________

Other possible requirements:

1. ________________________________
2. ________________________________
3. ________________________________
4. ________________________________

Possible problems:

1. ________________________________
2. ________________________________
3. ________________________________

Permit fees ____________________

Length of Permit ____________________

Departments involved in obtaining a permit:

HEALTH DEPARTMENT

Person contacted ____________________

Person needed to sign permit ____________________

Location ____________________

Phone ____________________

Time factor till completion ____________________

Requirements:

1. ____________________
2. ____________________
3. ____________________

Other possible requirements:

1. ____________________
2. ____________________
3. ____________________
4. ____________________

Possible problems:

1. ____________________
2. ____________________
3. ____________________

Permit fees ____________________

Length of Permit ____________________

Departments involved in obtaining a permit:

FIRE DISTRICT

Person contacted ______________________________

Person needed to sign permit ______________________________

Location ______________________________

Phone ______________

Time factor till completion ______________________________
Requirements:

1. ______________________________
2. ______________________________
3. ______________________________

Other possible requirements:

1. ______________________________
2. ______________________________
3. ______________________________
4. ______________________________

Possible problems:

1. ______________________________
2. ______________________________
3. ______________________________

Permit fees ______________

Length of Permit ______________

Departments involved in obtaining a permit:

ENGINEERING DEPARTMENT

Person contacted ____________________

Person needed to sign permit ____________________

Location ____________________

Phone ____________________

Time factor till completion ____________________

Requirements:

1. ____________________
2. ____________________
3. ____________________

Other possible requirements:

1. ____________________
2. ____________________
3. ____________________
4. ____________________

Possible problems:

1. ____________________
2. ____________________
3. ____________________

Permit fees ____________________

Length of Permit ____________________

Departments involved in obtaining a permit:

FLOOD CONTROL DEPARTMENT

Person contacted ____________________

Person needed to sign permit ____________________

Location ____________________

Phone ____________________

Time factor till completion ____________________

Requirements:

1. ____________________
2. ____________________
3. ____________________

Other possible requirements:

1. ____________________
2. ____________________
3. ____________________
4. ____________________

Possible problems:

1. ____________________
2. ____________________
3. ____________________

Permit fees ____________________

Length of Permit ____________________

Notes

1. Just because you are told you can't do something a certain way by the building or health department, do not take that for a final no. Ask to see what regulation prohibits you from doing what you plan and read it carefully to see whether it is actually prohibitive or was misinterpreted. If it is indeed prohibitive and you want to proceed anyway, find out what the variance procedure is. At each stage, try to anticipate what the response will be to your request and know what your response or next move will be. Thoroughly investigate the facts and present your case well. Anticipate the governing bodies' objections and have reasonable answers prepared. The most essential thing is perseverance—they may just tire of you and tell you to go ahead. I have become such a pest at times that the building department realized that the only way to get rid of me was to let me have my way.

2. Most areas regulate the stage of completion a house must reach before you are allowed to move in. Differentiation between contractor-built homes and owner-built or owner-contracted homes is common. In many areas of California, for instance, no one can move into a home built by a general contractor until everything is finished and the final inspection has been completed. If an owner is building or contracting their own home they are often allowed to move in sooner. If you have a finished bedroom, bathroom, and kitchen you are often allowed to move in. Also, many inspectors will look the other way if you are living in the house, being sympathetic to your situation. It is a difficult law to enforce.

But, even if you are allowed to move in it does not mean that you will be given your final inspection until the house is finished. This can have a few ramifications. You will not be able to get your long-term mortgage loan until the final inspection has been completed. This may cause you to stay longer on your construction loan at a higher interest rate, or worse—your construction loan may become due before you are able to get your mortgage loan. This can be disastrous.

3. Often the tax base for houses under construction is somewhat higher than after they have been completed (passed final inspection). If you take two years to build your house you may have to pay the higher tax base until then.

4. The Uniform Building Code states you must begin work within nine months of receiving the permit. Most all other codes only allow six months.

5. The penalties for not obtaining building permits are not that strong—you won't go to jail. But there are repercussions. You may put yourself in poor standing with your local permit department and if you are discovered they can red tag (a cessation of all work notice) your house until you get a permit. This delay can drag out for some time. Also there will be a monetary penalty for not having gotten a permit to begin with. In many areas the permit fees will be doubled if you begin work without a permit. This applies especially to areas covered by the UBC. Though other codes, such as the BBC, do not call for a penalty, local jurisdictions may. [Also, there are other reasons you may have to get the permit. Electricity companies may not hook up your temporary construction lines until they have proof of a permit.]

6. Sometimes different codes and regulations apply to different types of structures. For instance, if you call your guest house a studio it may not be required to have a bath and a kitchen. If it is called an agricultural building it may not have to comply with codes at all. Also, the tax base may be lower. If upstair rooms or lofts are called storage areas instead of bedrooms, ladders may be used instead of full stairs, and the ceiling can be lower. You may be able to install a smaller septic system since its size is based on the number of bedrooms and bathrooms. Sometimes the taxes are based on the number of bathrooms and bedrooms. Before you label your plans and turn them in to the building department, you may want to check this out.

7. Usually the local unions will have no problems with someone doing their own work. If, however, you are hiring non-union people to do the work for you and you are in a strongly unionized area you may get a little static, especially if construction starts are low and there are a lot of union carpenters hanging around the hall getting paid to picket non-union jobs. All of this is rare so don't worry about it unless there is a history of problems in your area.

Additional Information

8/Codes and Inspections

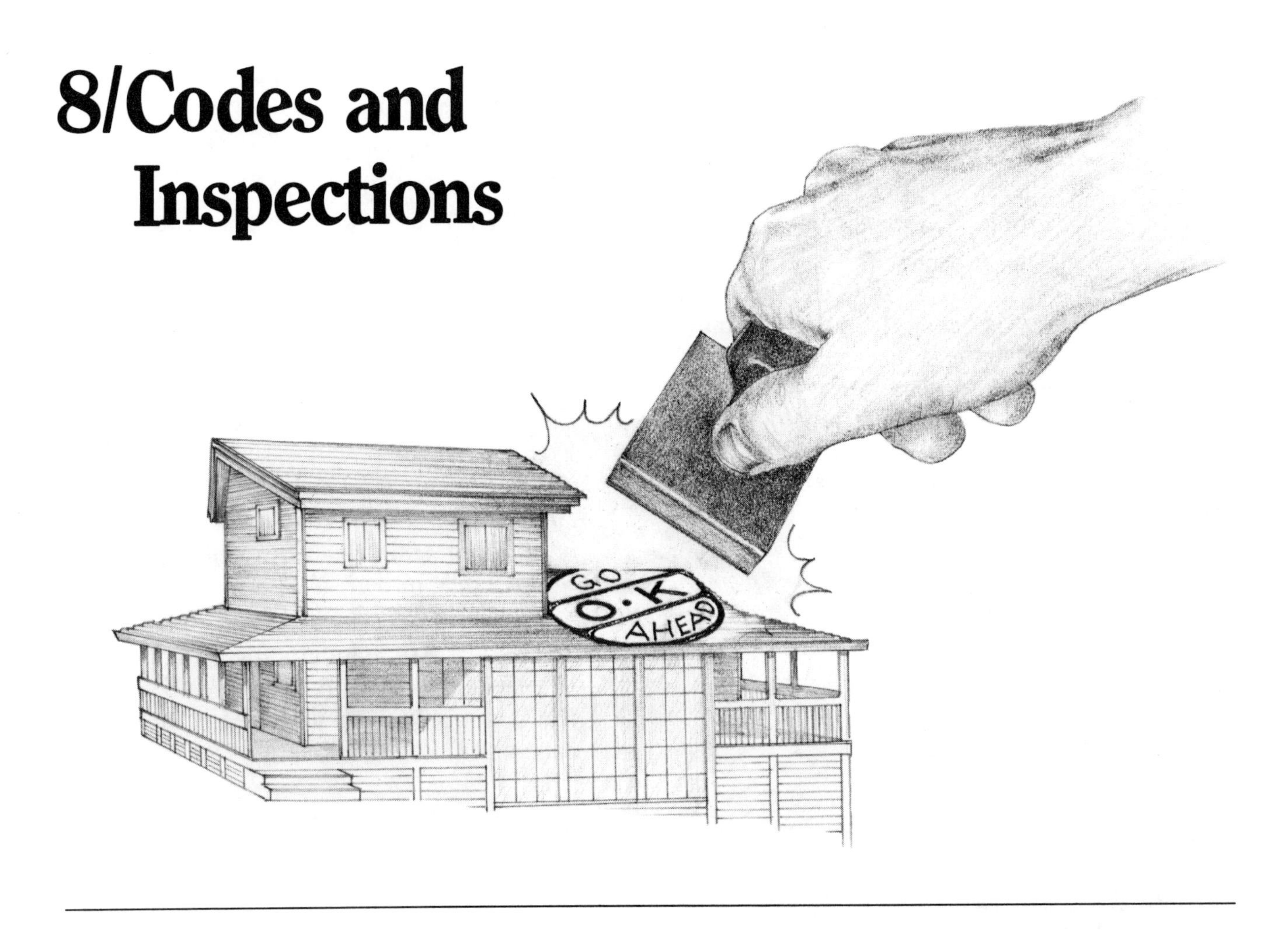

The provisions of this code are not intended to prevent the use of any material or method of construction not specifically prescribed by this code, providing any such alternate has been approved.
—Section 108
One and Two Family Dwelling Code

IN MANY AREAS codes and inspections are a fact of life and you will have to learn to live with them. They serve a purpose, but pressure from special interest groups has caused many of them to veer from their original intention. The majority are responsible and have the benefit and safety of the homeowner in mind. With all the other challenges in housebuilding, it is usually best to learn the codes and abide by them rather than fighting to change them unless absolutely necessary. Perhaps soon some sort of coordinated action can be applied to the building officials to ease codes where health and safety do not apply. A recent owner-builder code adopted in some counties of California could be precedent for that, but for the present it is best to assume you will have to abide by them, if they exist in your area.

Do not be intimidated by the code book; it is written in English and decipherable, though sometimes effort is required. Whenever you do not understand something about the code, ask your field inspector. Also, it is wise at every inspection to ask your inspector if there are any problem areas, as far as code enforcement is concerned, that you will be doing before the next inspection. Whenever you are in doubt, call and ask the field inspector (as opposed to the inspector behind the desk) how they would like to see it done. At each phase of construction be sure you completely understand the codes, *before* you begin that phase. Ask local builders, owner-builders and the inspector about any problem areas. I once had to jack up a 2200 sq. ft. floor system because we were ignorant of a code concerning the foundation knee wall.

Each inspector is different; some can be of great help to you, others can be a source of trouble and worry. Try to enlist them as your advisors and ask them for their support in your project. They are not the enemy. At the same time do not threaten the inspector's job by expecting them to pass poor workmanship or non-code and non-approved building practices. Also, show them that you know the codes or are willing to learn them and, though not experienced, you do your work as a professional. Unless it is necessary, do not ask for special privileges. If you want to veer from the codes, that is allowed through a process of variances and special approval but is usually not in the jurisdiction of the field inspector. In any case establish a good rapport with your inspector; you will both benefit from the relationship.

For Further Reading

Locally Enforced Building Code Book:

The Basic Building Code. BOCA International, 1313 East 60th Street, Chicago, IL 60637.

One and Two Family Dwelling Code. Council of American Building Officials, 560 Georgetown Building, 2233 Wisconsin Avenue, Washington, D.C. 20007.

The Standard Building Code. Southern Building Code Congress International, Inc., 3617 Eighth Avenue, South Birmingham, AL 35222.

Uniform Building Code. International Conference of Building Officials, 5360 South Workman Mill Road, Whittier, CA 90601.

The Building Code Burden. Field and Rivken. Lexington, MA: Lexington Books.

The Owner Builder and The Code. Ken Kern, Ted Kogon and Rob Thallon. Owner Builder Publications, PO Box 550, Oakhurst, CA 93664.

Codes and Inspections

What national code system applies in your area?[1]

- ☐ UBC 1976
- ☐ UBC 1979
- ☐ Standard Building Code
- ☐ Southern Building Code
- ☐ Basic Building Code
- ☐ Other ____________________

Are there additional local codes? ☐ YES ☐ NO

If yes, have you obtained these to be sure they do not interfere with your proposed project? ☐ YES ☐ NO

Are state energy codes in effect?[2] ☐ YES ☐ NO

____

____

____

____

Will the state energy codes limit your proposed window schedule? □ YES □ NO

Are there any other limitations? ______________________________

____

____

Who will do energy calculations on your home, if required?

Name ______________________________
Cost ____________________

Will codes allow wood stoves to be used as main source of heat? □ YES □ NO

Have you studied the solar and energy efficiency tax credits in your area to see what might apply to your house?[3] □ YES □ NO

____

____

____

What is/are the inspection requirement(s) in your area?[4] □ YES □ NO

____

____

____

____

____

____

____

When are inspectors available to visit site?

- □ Monday
- □ Tuesday
- □ Wednesday
- □ Thursday
- □ Friday

Is one inspector assigned to your area? □ YES □ NO

If yes, name: ______________________________

Is there a reinspection charge?[5] □ YES □ NO

How often will the inspector come out? ____________________

How much notice is needed before an inspection? ____________________

What kind of inspections are needed?

____

____

____

____

____

____

____

Are fire marshal inspections required? □ YES □ NO

Is there a maximum time between inspections? ____________________

Will any work be required to be done by a licensed person? □ YES □ NO

If so, what?[6] ____________________

____

____

What codes are in effect?

- □ Electrical
- □ Fire
- □ Plumbing
- □ Building
- □ Mechanical
- □ Swimming pool

Will local codes allow you to do your own:

- □ Design
- □ Engineering
- □ Plan drawing
- □ Contracting
- □ Electricity
- □ Plumbing
- □ Carpentry
- □ Grading

Do you have to register with the state as an employer and withhold or remit income tax, social security, etc.?[7] □ YES □ NO

Is there a waiver available?[7] □ YES □ NO

Can you build to sell or rent and still do your own work? □ YES □ NO

Are you required to list your subcontractors on permits?[8] □ YES □ NO

Is temporary occupancy before the home is completed allowed in your area? □ YES □ NO

Are codes more lenient for studios, garages, agricultural building, storage sheds, etc.? □ YES □ NO

Are codes more lenient if you rename some of your rooms? □ YES □ NO

Will codes allow your type of construction and are you allowed to deviate from the code?[9] □ YES □ NO

Will you be able to live on your property (i.e., in a mobile home) while you are building? □ YES □ NO

If so, what requirements must you meet (septic, well, road, power, etc.)?

____

____

____

____

Can recycled materials be used?[10] □ YES □ NO

Can you cut your own lumber?[11] □ YES □ NO

Are carports or garages required? □ YES □ NO

How many off-street parking spaces are required?

Will areas where dirt is to be filled or dumped need to be inspected first? □ YES □ NO

Are temporary toilet facilities required during construction? □ YES □ NO

____

____

What are their monthly costs? □ YES □ NO

Do foundation layout lines need to be inspected for proper setback requirements prior to excavation?[12] □ YES □ NO

Notes

1. Rather than write their own codes, most building departments adopt one of the national building codes. To this they add any special regulations that they have also adopted. For the most part, all of the national building codes are the same and are modified to fit special conditions in the area of the country that they serve. Ask your local building department if any codes are in effect and, if they are, get a copy of the code book and local code regulations. Believe it or not, code books are written in English and can be deciphered with enough time and effort. They are not written with the novice in mind and can sometimes be difficult to understand.

2. State energy codes are now becoming more prevalent. They stipulate certain ways in which a house can be built to insure that it does not use or waste too much energy by regulating the amount of window space, the heating sources, weather stripping, and similar features. Sometimes calculations are required to show the amount of energy the house will use and lose. These can be calculated by the owner but sometimes it may be best to hire a local professional, if the cost is not too high.

3. The federal government and many states offer liberal tax credits in order to encourage people to use solar energy and other energy efficient techniques. These credits are direct deductions from your taxes and can be applied to both active and passive solar systems. It is wise to get well acquainted with the credits before you begin to design your house since this knowledge may aid you in designing to get the most of both energy efficiency and tax credits.

4. Not all areas of the country have building inspections. However, those which don't are becoming fewer all the time. I remember building in North Carolina in 1969 and the only inspection required was a septic tank inspection. In most suburban, many rural, and all urban areas today, the work must be inspected as it progresses to be sure that it meets specific standards to insure quality. Here is a typical sequence of inspections from septic tank installation to the final inspection for a house in the San Francisco Bay Area of California, one of the tougher areas to build in as far as inspections (and most everything else, for that matter). Your own area may differ.

Number	*Inspection*
1	Land where any fill is to be dumped is inspected.
2	Water is tested for flow and purity.
3	Soil is inspected for its ability to perc.
4	Septic tank is inspected as it is installed.
5	Any excavating that is to be done is inspected.
6	After foundation form boards are in place and rebar is set, the forms are checked.
7	Foundation is poured, and floor frame is built, rough plumbing and electricity and heating ducts are installed below the floor surface, and before subflooring goes on, the underfloor utilities are inspected.
8	The house is framed but before wall and roof sheathing is covered the nailing pattern is inspected. Also the above floor rough plumbing and electricity and heating ducts are installed and inspected.
9	The outside of the house is finished, insulation is installed and inspected.
10	Sheetrock is nailed up and nailing pattern is inspected before it is covered.
11	FINAL INSPECTION—house is now completed.

5. If the inspector is visiting your site and finds something wrong and then has to return, there may be a charge to you for the return trip. You are usually given a free chance and if they have to return again after the first notification there may be a charge. In many areas they keep checking until you do it right and there is never a charge. Often you can just ask the inspector to drop by to ask them about a situation.

6. If you are building a house to live in yourself, you

can do most or all of your own work. In some areas a licensed electrician or plumber may be required. If you are building a house to sell on speculation, or for someone else, you may not be allowed to do the actual work. In California, for instance, you are only allowed to do all the work yourself if you are planning to live in the house for a year. Find out what your local regulations are before beginning. If licensed people are required for parts of the job, you may have a friend who is licensed who will sign for you as you do all the work (essentially you are acting as their employee).

7. If you are building or contracting your own house, you are now in the role of an employer. As such, there are certain requirements you may need to fulfill, such as payroll taxes, registering with the state as an employer, getting an employer's number, etc. It is wise to do this if it is required as there could be complications later if you fail to do so. These requirements can sometimes be waivered if you're only employing people for a limited time.

8. Some areas may require you to list your different subcontractors on your permit and their license numbers if you are not doing all the work yourself. This helps in preventing unlicensed workmen from working as contractors. If that is a problem, and you are allowed to do all the work yourself, you could just list yourself and then hire unlicensed workers.

9. Be aware that there is a section of the building code that essentially says you can do whatever you want if you can prove to the building officials that it is structurally sound. Section R-108 of The One and Two Family Dwelling Code reads:

> The provisions of this code are not intended to prevent the use of any material or method of construction not specifically prescribed by this Code, providing any such alternate has been approved.
>
> The building official may approve any such alternate provided he finds that the proposed design is satisfactory and complies with accepted design criteria.
>
> The building official may require that evidence or proof be submitted to substantiate any claims that may be made regarding its use.

The Preface of the Basic Building Code reads:

> By presenting the purposes to be accomplished rather than the method to be followed, the Basic Building Code allows the designer the widest possible freedom and does not hamper the development. It accepts nationally recognized standards as the criteria for evaluation of minimum safe practices, and for determining the performance of materials and systems of construction.

And finally, from the founder of the Building Officials and Code Administrators of America (BOCA), Rudolf Miller:

> The building laws should provide only for such requirements with respect to building construction and closely related matters as are absolutely necessary for the protections of persons who have no voice in the manner of construction or the arrangement of buildings with which they involuntarily come in contact. Thus, when buildings are comparatively small, are far apart, and their use is limited to their owners and builders of them, so that, in case of failure of any kind they are not a source of danger to others, no necessity for building restrictions would exist.

10. You may be able to find used or recycled materials at a much better price. In the case of lumber it may be a better grade of wood since it is dry and may have been cut from larger trees than are now available. Ungraded materials can be incorporated throughout the house, with the exception of the frame. By code, framing materials must be graded. If you are planning used lumber for the frame of the house, the inspectors may say nothing if they recognize the quality to be high. They may, however, ask that you either get it graded by a local lumber grader (ask at your local lumber mill), or okayed by a licensed structural engineer or architect. If the previous lumber grading stamp is still visible on the lumber, that too may suffice.

11. A number of different tools are now available which allow one or two people to cut rough lumber from logs on the site. The work is slow, hard, and loud. The product is not a finished, planed piece of lumber, but rough-cut. These then can be used as they are (they are stronger than finished lumber), but sometimes their irregularities can make finishing work a little more difficult. You can also take the lumber to a finishing mill after you have cut it and, at a very reasonable price, have it milled. Alaskan Mill is the name of one of the more popular one- and two-person saws.

12. Local governing bodies or the subdivision office will sometimes require you to lay out the perimeters of the building exactly on the lot using strings. They will then check these against your property lines (which also must be laid out in string) to be sure that you have the proper setbacks before any work or excavation can begin.

Addresses of Building Officials

1. West—*Uniform Building Code*

   International Conference of Building Officials
   5360 South Workman Mill Road
   Whittier, CA 90601

2. East and Midwest—*Basic Building Code*

   The Building Officials and Code Administrators International
   17926 S. Halsted Street
   Homewood, Illinois 60430

3. South—*Standard Building Code*

   The Southern Building Code Conference
   3617 Eighth Street South
   Birmingham, Alabama 35222

All three collaborated to produce:

One and Two Story Dwelling Code

Available: $10 from any one of the above or

Council of American Building Officials
560 Georgetown Building
2233 Wisconsin Avenue N.W.
Washington, D.C. 20077

Additional Information

Additional Information

9/Insurance

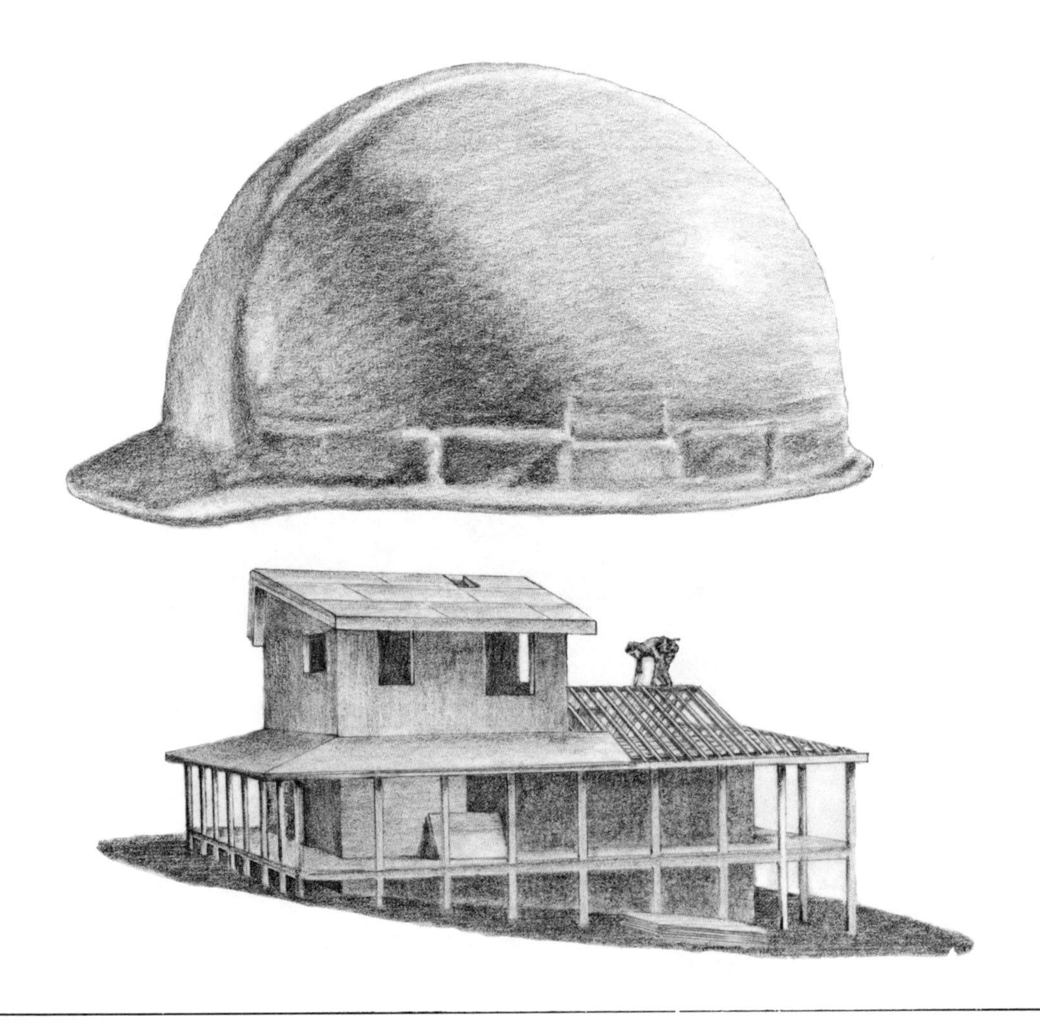

Construction work has its dangers. Insure yourself and workers well.

IT IS EASY for owner-builders, being a bolder element of the population to begin with, to overlook the need for certain types of insurance while building. My advice is: **DON'T!!** Many lenders will require you have certain types of coverage to safeguard their investments; others are optional but definitely recommended. Construction, by nature, is a dangerous trade. It is wise to insure yourself, your friends and family, as well as your paid workers with workman's compensation insurance. Fire insurance is also advised. Building or contracting your own house takes a lot of optimism and trust; healthy caution and insurance help fill out the package.

Insurance

Is workman's compensation insurance required or desired?[1] □ YES □ NO

Is the jobsite adequately protected by liability insurance?[2] □ YES □ NO

Are you insured against theft and vandalism and, if so, what are the conditions of the policy?[3] □ YES □ NO

____

____

____

____

____

____

Is life insurance required by the lender?[4] □ YES □ NO

Are you covering yourself with workman's compensation and life insurance?[5] □ YES □ NO

Are you covering friends and relatives, even those working for free or trade, with workman's compensation? □ YES □ NO

Will the house be adequately insured against fire and will this amount increase as work progresses? □ YES □ NO

Notes

1. Workman's compensation insurance covers people injured while working on your home. It not only pays for medical expenses, but for lost work and recuperation time as well. In many states it is mandatory and you cannot get your permit until you present proof that you have taken out a policy on the project. In any case, it is wise to have coverage for your workmen, yourself, your family, and friends. Construction is, by its very nature, a dangerous job and one injury can cost a fortune in hospital and doctor bills and lost work. If you have workman's compensation for everyone working on the site, both paid and unpaid, you are protecting them in the event that something should happen. If you do not have workman's compensation and something does happen, there is a good chance that you can be sued and will lose because it is negligent of you not to have the workpeople covered by workman's compensation.

Workman's compensation has a different base rate for each profession. In construction it is about 9% of each person's pay. That means if you are paying someone $10 an hour you have to pay an additional 90¢ per hour towards workman's compensation insurance. At the beginning of the job, you take out a policy by estimating how many dollars of wages you plan to pay. For friends or family working for free you bill their labor at the equivalent price you would have had to pay if they were earning wages. If the total of all of this comes to $10,000, for example, you then pay the insurance company $900. After the house is done, or a year has passed, you report back to the company with payroll sheets showing what you actually spent and adjustments are made.

2. Aside from protecting your workers it is wise to cover the jobsite with liability in case anyone who is not a worker should get hurt. This is a blanket policy that covers everything from delivery people to children who wander into the jobsite. It is worth the money.

3. Very often, theft policies apply only to materials that are actually attached to the house, not those stored in the yard waiting to be applied to the structure. For these it may be difficult or impossible to get insurance. If that is the case and you feel there is a possibility of theft you may want to buy only the materials that you need or store them in a protected storage area. It may also be wise to ask the neighbors and local police about keeping an eye on your project.

4. If you are borrowing money to build the house yourself, much depends on you. Because of this, lenders may require life insurance. Should you die, they would then have enough money to hire people to come in and finish the project.

5. You may want to cover yourself by both workman's compensation and life insurance. If something happens to you, money will then be available to finish the project or to help if you are injured. I would recommend you get both of these.

Additional Information

Additional Information

10/Financing

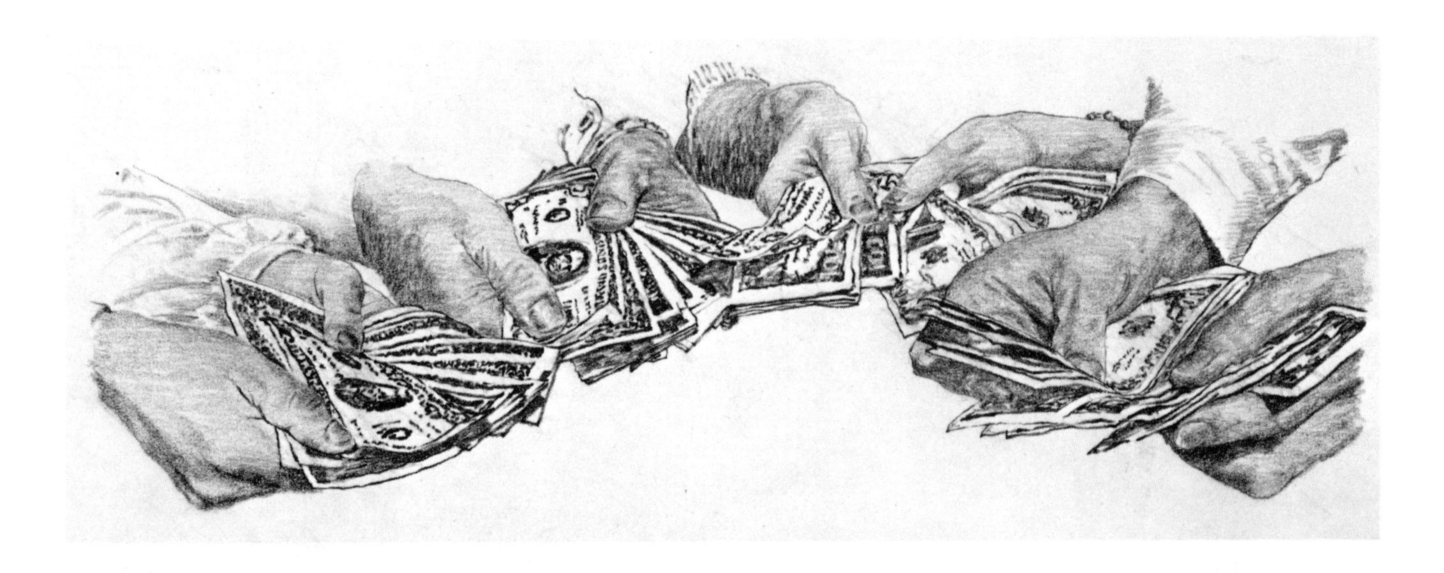

$70,000.00 borrowed at 14½% for 25 years costs $869.52 monthly. Over the life of the loan you will pay $260,856.00. $190,856.00 goes to the bank as interest. If you paid $869.52 each month directly towards your project, in 6.7 years you would have raised the needed $70,000.00 (no adjustment has been made for inflation). For the other 18.3 years you are paying for the "privilege" of borrowing the money all at once.

THE MOST DIFFICULT part of financing an owner-built or owner-contracted home is acquiring the money needed to finance the actual construction. This is what is called an "interim" or construction loan. Lenders do not know what kind of job you will do, hence their natural hesitancy in financing your project. Once the house is completed and the value can be appraised, you will find acquiring a "take out" or mortgage loan is a much simpler matter.

Given the lack of financing money for new homes, and the incredibly high rate of interest which causes a burdensome monthly payment almost all of which goes to the bank in interest, we recommend that you consider financing the construction of your home with your own money. This may be difficult and cause the construction to go even years slower, but when it is done the house is yours and no huge monthly payment is needed. How much peace of mind is created by your new home if you must struggle each month to pay for it?

If you do need to borrow a large amount to build with, consider asking your friends and family to loan you the money. You can secure their loans by a deed of trust on your lot and house issued in all the lenders names, and pay them at rates similar to those you would pay the bank for construction loans. This can be much simpler to deal with, as construction loans from lending institutions are incredibly difficult and problematic to administer. It also may be simpler to ask the people who believe in you to back the construction as opposed to the bank. Also it allows them to take advantage of interest rates they would not otherwise be able to get. Once the house is completed and you have received your mortgage, you can pay them back.

The following chapter was prepared by Neil Hickey. Neil is president of Plan-A-Visions, Inc., a financial planning and advisory company that specializes in assisting owner-builders' and owner-contractors' construction financing.

For Further Reading

Action Planning for the Owner Builder. Neil Hickey. Plan-A-Visions, 4414 Bacon Court, Pleasanton, CA 94566.

Financing

Where to find money sources or look for loan assistance:

1. Commercial banks
2. Savings & loan institutions
3. Life insurance companies
4. Credit unions
5. Construction financing companies
6. Construction financial planners
7. Mortgage bankers
8. Mortgage brokers
9. Loan negotiators
10. Financing companies
11. Real estate brokers
12. Friends and relatives
13. Finance land by self, use own cash for construction
14. Special government programs

Do you or your property qualify for special government financing programs?

Housing Urban Development (HUD) Programs? □ YES □ NO
Farmer's Home Administration Programs? □ YES □ NO
Veteran's Administration (VA) Programs? □ YES □ NO
Federal Housing Administration (FHA) Programs? □ YES □ NO
Special City-Sponsored Housing Programs? □ YES □ NO

Is your Credit Union a possibility? □ YES □ NO

Is it possible to obtain the money from other assets, such as real property, by a second mortgage, refinancing, or through the sale of the asset? □ YES □ NO

Do you have any friends or relatives who have the cash and who would loan you the money to build? □ YES □ NO

Potential lenders:

____

____

____

____

____

Do you have any friends or relatives who could get cash by using their assets as collateral for a loan and give the money to you? □ YES □ NO

Potential lenders:

____

____

____

____

____

Do you have any friends or relatives who would underwrite your loan as a cosigner? □ YES □ NO

Potential underwriters:

____

____

____

____

____

Do you have a skill or talent you can trade for construction labor and/or materials? □ YES □ NO

Is it possible for two or more families to qualify and live in the house? □ YES □ NO

Is it possible to live on the land in a temporary shelter (R.V., trailer, etc.) and apply the savings on your monthly rent or mortgage allotment towards buying materials? □ YES □ NO

Monthly allotment: ________________

Would it be possible to build a small "core" house and later on add more rooms with a home improvement loan, personal loan, or with your own cash? □ YES □ NO

If yes, what are the possibilities? __________________________

____

____

____

____

____

Does the building material supplier have a financing program? □ YES □ NO

Is it possible to use salvaged and recycled materials, or to perhaps decrease the size of the house and finance it yourself? □ YES □ NO

Are there other areas of the state or country where owner-building or owner-contracting may be simpler? □ YES □ NO

If yes, where: ____

____

____

If you are considering a prefabricated home, does the prefab company offer financing? □ YES □ NO

Will the seller of the land consider giving you a construction loan? □ YES □ NO

Will the seller of the land subordinate to a construction loan?[1] □ YES □ NO

Will the construction lender grant a lot advance in the construction loan?[2] □ YES □ NO

Will you and your property qualify for a construction loan through a bank or savings and loan association? □ YES □ NO

Selecting a conventional lender for your project:

Will the bank make a loan directly to you? ☐ YES ☐ NO

If yes, Name of Bank: ______________________

Loan Manager: ______________________

Will loan be for materials only, or will you have to borrow the amount for a contracted home?

☐ Total: ____________
☐ Materials only: ____________

Will the bank require license number of a licensed contractor? ☐ YES ☐ NO

If yes, Contractor: ______________________
License number: ______________________

Will a building control company be required by the bank? ☐ YES ☐ NO

If yes, Name of company: ______________________
Address: ______________________
Phone: ____________
Fee: ____________

Will the lender require the loan amount to be as high as if a professional contractor were doing the building? □ YES □ NO

Will the lender require a reserve account in the cost estimate?[3] □ YES □ NO

If so, how much? ________________

Does the lender have a maximum loan limit? □ YES □ NO

Will the lender finance only primary homes which are owner-occupied? □ YES □ NO

What monthly income ratio does the lender use to qualify the borrower for the loan amount?

Ratio: ________________________

If the property is not paid in full will the lender consider:

□ Subordination of the existing mortgages on the land?[1]
□ Giving a lot advance in the construction loan?[2]

What is the time allowed for construction? ________________________

____

How is the money made available during the construction period? ________________

____

____

Is a voucher system used?[4] □ YES □ NO

What is the repayment schedule of the construction loan?

____

____

What is the interest rate of the construction loan? ________________________

How is the interest calculated during construction? ________________________

How many points will be charged for the loan?[5] ________________________
Can the points be financed in the loan? □ YES □ NO

Is there any type of prepayment penalty? □ YES □ NO

If yes, what? ________________________________

Does the construction lender include or promise a permanent long-term real estate loan? □ YES □ NO

What are the conditions of the permanent loan?

Rate? ____________________, Fixed or Variable? ____________________

Must the rate be negotiated during the life of the loan? □ YES □ NO

What are the points charged for the permanent loan? ____________________

What is the term? □ 20 years □ 25 years □ 30 years

Is the loan assumable? □ YES □ NO

Is there a prepayment penalty? □ YES □ NO

Do the monthly payments remain the same during the life of the loan? □ YES □ NO

Will a performance bond be required?[6] □ YES □ NO

Can any construction work be started prior to obtaining the construction loan? □ YES □ NO

How long does it usually take to obtain the loan? ____________________

What documents and exhibits are required by the lender to process the loan request?

For credit verification? ____________________
For income verification? ____________________
Property documents? ____________________
Construction documents? ____________________

How much of the house must be completed before a long-term mortgage loan is possible?

____

Are any of the following needed by the bank before they will lend you money?

- □ Title search
- □ Fire insurance
- □ Life insurance
- □ Building permit
- □ Subordination clause
- □ Contractor's license number
- □ Control company

Notes

1. *A subordination agreement* is an agreement which provides that, under certain conditions, an existing lien (mortgage) may change position and become a junior lien to a new lien.

Example: The holder of the first mortgage on the land agrees to change his first mortgage position to a second mortgage position to enable the construction lender to assume the first mortgage position.

2. *A lot advance* is a situation where the construction lender advances the money in the construction loan to pay off the property.

3. *A reserve account* consists of extra funds set aside from the construction loan and is normally not available until the structure is completed. The amount is usually 10–15% of the projected costs for construction.

4. *A voucher system* is a system ordering payment for construction services and materials if the payment is legitimate and there are sufficient funds to cover the voucher amount. The voucher is like a check. The maker orders the payment and the agency who issued the vouchers will send a check directly to the payee.

5. A point is calculated at 1% of the loan amount. Example: 3 points on a $40,000 loan would be calculated as 3% of $40,000, which is $1,200.

6. A bond guaranteeing that the project will be completed and the bills paid is a performance bond.

Additional Information

Additional Information

11/Estimating

How often do you hear of houses costing more than their estimated costs? How often less?

READING AND TRULY comprehending this chapter may be one of the most important things that you do in your life. I am not saying that just to get your attention, but because it is true as well. It may save you from a dread disease infecting almost every owner-builder known and most of the building industry as a whole. I myself have fallen victim to its agonies and heartbreaks more times than I like to remember. I have seen its ravages on pocketbooks and lifestyles, seen the damage done to family unity and human psyche from North Carolina to California. In case you have not yet guessed what this dread disease is, I will tell you: UNDERESTIMATION. If this book does nothing more for you than save you from that disease, it has paid for itself a thousandfold. Read the following guide and then fill out the workbook as objectively and honestly as you can. Then add a 20% "fudge factor" and you can probably feel safe.

Everyperson's Guide to the Perils of Estimating

The house which actually costs less than its estimate is so rare that it ascends into the realm of miracles. Houses go over their estimate so commonly that it seems inevitable. The consequences of projected costs exceeding the amount planned can be ruinous—an additional several hundred dollars a month in loan or mortgage payments, for as long as thirty years. Such a long-term burden can seriously limit your ability to realize other fantasies, like vacations, a new car or boat, or travel, for twenty to thirty years. Before you take this risk, consider some things which may put you ahead in the art of estimating.

In estimating housebuilding costs, our fallibility is usually greatest in two areas. First, we try to incorporate more features into our designs than we can afford. Second, adequate account is not made for unknowns in the job—the hidden costs.

For those of us who feel we are putting up the only house we will ever get to build, and one we may live in for many years, desires to have everything we want will tempt us with peculiar potency. The idea of cutting back on features we have dreamed of for many years appalls us, but changes along the way which seem small may add up to a large increase in price. The tendency is to choose designs which are larger, more complex, or more expensive than those which we can actually afford.

It pays to remember the scale of your investment. In planning a house we make decisions involving thousands or ten-thousands of dollars. Most of us lack familiarity with spending these amounts and our perspectives can be lost, and regained too late. Give yourself a healthy margin of estimating. Allow some decisions to be made later in the process; as construction progresses, final costs become more predictable. That may be the time to decide whether to add an extra wing, what furnishings and fixtures to use, or how big the garage will be. Always try to remain conscious of how much money you can afford to spend and of how much house that money will buy.

There are aspects of the art of estimation which make it tricky, even for a pro. (The contracting business casualty rate ranks it with the riskiest, and it is poor estimating which often boots otherwise competent people out the door.) Very accurate estimates can be made on conventional or tract homes. Most of these homes are designed to use prefabricated components such as plywood, paneling, sheetrock, and cabinets. These go up with predictable speed and energy input. Also, since similar houses go up regularly, there are relatively few unknowns. Work can progress smoothly and cost calculations can be based on square footage. Any knowledgeable person in the housebuilding industry in your area should be able to give you a reliable cost factor to multiply by square footage.

Custom building, on the other hand, challenges any estimator. Many builders will operate only on a cost-plus basis since any custom house is unique, a prototype. One like it has never been built before and the problems are unknown. Usually such structures do not lend themselves to the use of premanufactured materials or components. Therefore, these houses are very labor-intensive and expensive. I am not recommending that you forget custom building, for to my mind custom homes often possess more life and character than conventional ones. However, be aware of their added costs!

Non-house expenses such as wells, roads, septic systems, water storage and filtering systems, and getting power to the land often overshadow the cost of construction even when the basic estimate of the structure itself remains sound. Remember that there are often legal fees required, and custom building may necessitate costly structural engineering fees, also.

Inflation often leads to underestimation. Housebuilding costs for materials and labor may rise 5–10% just in the six months between the drawings and calculation of the estimate, and the start of work. Similar changes which will affect your estimate can occur during the construction period. Building materials can jump 10–30% overnight. Allow margins for these raises and a 12–18% rate of inflation per year.

The site and location of a house can bear strongly on its costs. Costs rise with difficult or inaccessible siting. Foundations may need to be more complex on steep slopes, and construction is generally more difficult. Locations remote from hardware and building supply stores make the job more time consuming for the contractor, carpenters, and other workmen, and add to costs. Delays and added expense in the delivery of materials can frequently occur in remote areas. Also, power, roads, and water in such instances can become prohibitively expensive.

Ineffective organization can be costly. The entire project requires a considerable amount of coordination between people. Poor organization results in less production. If contractors and workmen slip even 10–20% from maximum efficiency, thousands of your dollars may be wasted. Try to insure that all the people involved work together well and are interested in providing maximum productivity for minimum time and effort.

Remember, housebuilding is a many-faceted undertaking. Estimating costs is a key process and can be a difficult one. You will have a good headstart if you avoid some of the pitfalls I've been discussing. Watch for sudden changes in prices and hidden costs. Remember there is, at times, a tendency in each of us to be somewhat too optimistic. We estimate on the low side, in an effort to convince ourselves that our dream house can be all that we want it to be. Look at whether you are doing that and remind yourself to take care and be cautious with your dollars.

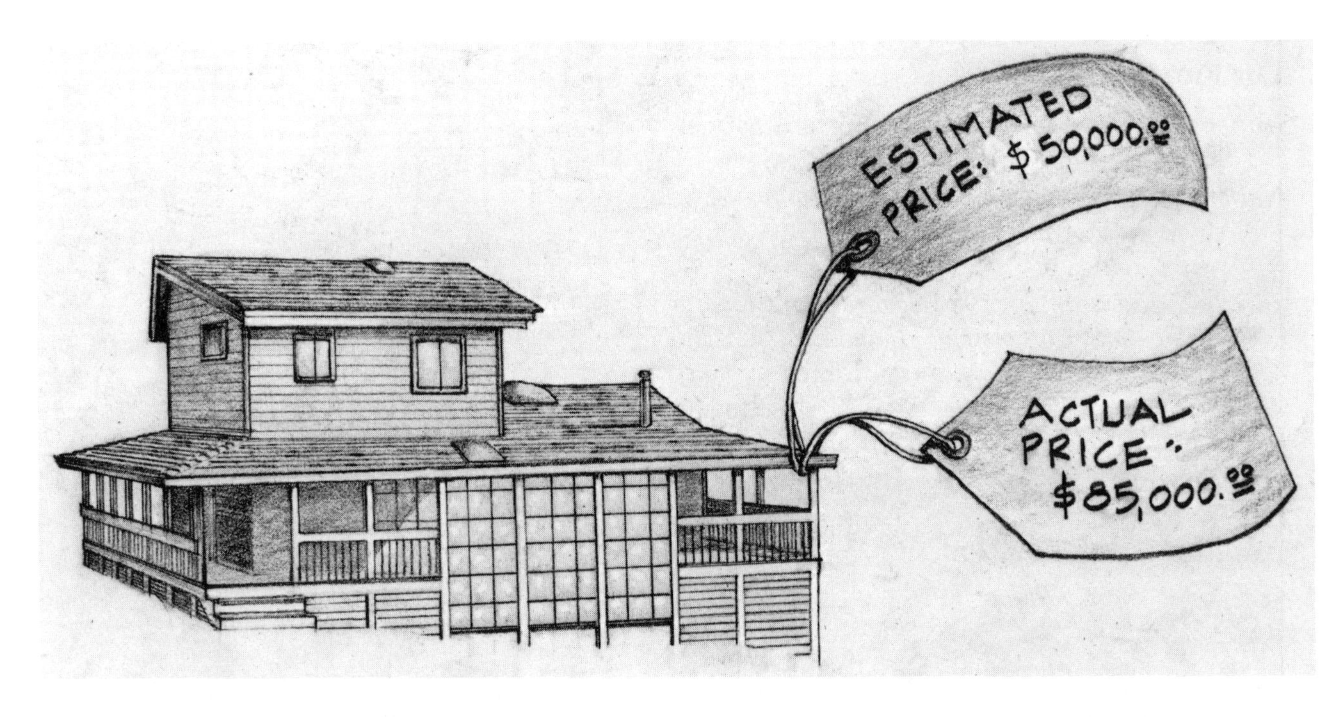

Estimating Guide

	EXAMPLE HOUSE	ESTIMATED COST	FINAL COST
PRECONSTRUCTION:			
Legal Fees for buying land, handling contracts, etc.	$157.50	______	______
Design and Drafting	$1637.48	______	______
Permits	$695.50	______	______
Water Hookup	$250.00	______	______
Recording of Deed of Trust	$19.00	______	______
Insurance—Liability, Theft, Fire	$750.00	______	______
Engineering	$50.00	______	______
Portable Toilet (1 year)	$420.00	______	______
Land	$30,000.00	______	______
Closing Costs for Land Purchase	$250.00	______	______
Sewer Hookup or Septic	$250.00	______	______
Temporary Power	$167.50	______	______
Site Phone Hookup and 1 Year Service	$720.00	______	______
Roads, Bridges, and Driveways	$1950.00	______	______
Interest on Construction Loan (1 year loan at 18% for $75,000.00)	$13,500.00	______	______
Tools	$2160.00	______	______
Vehicle Expenses (at 19¢ per mile)	$2842.00	______	______
Solar Consultation	$300.00	______	______
Books and Classes	$300.00	______	______
TOTAL	$56,418.98	______	______

We have now spent about $56,500.00 and we are ready to begin construction. Your own costs may vary greatly, depending on your land costs and road costs and whether you need to dig a well and install a septic system.

For Further Reading

National Construction Estimator. Edward Sarviel. Solana Beach, CA: Craftsman Book Co.

National Repair and Remodeling Estimator, 1981. Albert S. Paxton. Solana Beach, CA: Craftsman Book Co.

Building Cost Manual. 1980 Square Foot Costs for Residential, Commercial, Industrial, Agricultural and Military Structures. Edited by Gary Moselle. Solana Beach, CA: Craftsman Book Co.

Additional Information

12/Inner Resources

If in building the house, you have adversely affected yourself, your loved ones or your relationship with them, has it been successful?

BUILDING OR contracting your own home is a major event which should never be entered into without a good understanding of what is involved and what will be demanded of you. It is a process that, once begun, must be finished—there is no turning back. Perhaps the scope of this book, which covers only the work to be done before you begin construction, gives you some sense of the size of the task. It is not a small one.

On a personal level, three things are important

to know about yourself before you begin: that the timing is correct, that you have the necessary emotional and spiritual resources, and that your commitment is strong enough to complete the project without tremendous strain.

As far as timing, you need to consider all the other factors that are in your life—jobs, children, careers, your marriage—and be sure that you can begin another major project without doing permanent damage to your present commitments. You need to be sure that you will have the love and support of the people close to you. You will need this, perhaps more than anything else. If you are building with a mate, be sure your relationship is healthy and strong before you begin, as the strain can be the final blow in a faltering relationship. And finally, be sure your commitment is total and deep, since all your inner resources will be needed and you must know there was nothing else that you really wanted to do but build this house.

The intention of this chapter is to ask you some questions that will require that you be very honest with yourself. I have left no room for answers as the answers should come from, and remain, within. Do not be satisfied with your first response; we often have set responses that are not really true. Look deep and be honest. If, after going through this chapter, your resolve is still strong and you feel you are ready to begin, jump in with no regrets or hesitations. You are about to begin one of the most rewarding and enlightening experiences of your life. Good luck!

For Further Reading

Illusions. Richard Bach. New York: Delacorte Press, 1978.

Lazy Man's Guide to Enlightenment. Thaddeus Golas. New York: Bantam Books, 1980.

Pattern Language: Towns, Buildings, Construction. Christopher Alexander, et al. New York: Oxford University Press, 1977.

The Work Book: The Politics of Building Your Own Home. Kern and Turner. Owner Builder Publications, PO Box 550, Oakhurst, CA 93664.

Inner Resources

Will your approach to this project be a professional one?

Do you really want to do this?

Do your children and mate also want to move and build this house?

Do you like this type of work on an everyday basis?

Will you enjoy the process of building your own home?

Will the weather be a problem?

Do you like working alone?

Do you resist doing things that you don't want to do?

Can you work in physically uncomfortable situations for long periods of time?

Do you begin with enthusiasm and then lose interest?

Do you take advice easily?

What are the other alternatives to building the house yourself?

How much money are you saving by doing all or part of the work yourself?

Where did the idea of building or contracting your own home come from?

What is the purpose for building or contracting your own home?

What are your goals in doing this project?

Are your motivations strong enough to see you through to the end in a good frame of mind?

Are you being realistic in your:

Time schedule?
Estimate?
Energy estimate?
Motivations?

What other projects have you never finished? Why?

Will this project be different? Why?

Do you need a lot of support and acknowledgement?

Does it bother you to progress slowly?

Are you careless about organization and administration?

Do you work well with people?

Are you willing to find out the best way to do something before proceeding?

Are you good at coordinating different peoples' energies?

What kind of physical condition are you in?

Who are the people that will be advising you and do they know the best way to do things?

Can you *afford* the house that you want?

Do you have trouble making decisions?

Do you like problem solving?

Are you emotionally prepared for this experience?

Can you build or contract this home and not let it seriously affect your close relationships?

Do you have positive ways of releasing tension?

How far away from the site will you be living?

Is there anything more you can do to prepare yourself?

Are you accident prone?

Do you need to consider changing your diet in any way to accommodate the new work load?

Should you consider building the house in stages?

Do you handle mistakes, both yours and others', well?

Do you know how to pace yourself well so you do not get fatigued?

Are you a fast, medium, or slow worker?

Do you know what design you really want?

Have you really allowed yourself enough planning, design, and preparation time?

Can you delegate jobs and authority well?

Are you capable of healthy self-discipline?

Are you very efficient?

If you are building with a mate, do you make decisions well as a couple?

Do you worry too much?

Is your design too unique and complicated for your time schedule and skill level?

Are you prepared to trust your own choices?

Do you have the basic confidence to undertake this?

Are you a good bookkeeper?

Will the neighbors resent your design?

Are you willing to do a lot of things you don't want to do?

When is the best time to start the project?

Are the goals and visions mutually shared by those involved?

Will patterns of dominance and submission cause problems during the construction?

In your partnership, or marriage, are both people able to get what they need or want without one overriding the other?

Should you become tool-efficient before beginning?

Are you willing to step back from a project to see its totality and look for easier ways?

Who will be the project manager and how much time will that involve?

Who will be responsible for gathering information?

How long are you willing for the project to take?

Who will complete the project if you are disabled?

13/Types of Construction

MANY NEW BUILDERS do not realize that there are many different types of construction they can choose from. Most builders proceed building the standard stud-constructed house, never stopping to consider their other choices. This chapter will explain the most common types of construction methods used in the United States today. Your choice in the type of construction you will use should be well thought out. Your decision will impact your cost, your design, the feel of your home and your building process. There should be definite reasons for choosing one type of construction over another.

Along with your decision on the type of construction you will use, you also will need to decide whether to use a pre-fab kit home or build the entire house yourself at the site (a site-built home). This is a very important decision too; it can greatly impact your building experience. Let's discuss this building issue first.

The Site-Built Home

Until a few decades ago, almost all houses were site-built homes (except for those ordered from the Sears catalogue). Most owner builders have avoided factory-built homes, feeling that they could not get the design or quality they wanted from a kit. This is really no longer the case. You still may, however, choose to build the entire home yourself for several reasons. Needless to say, you do have more control over your home should you decide to build at the site. You can choose and inspect every piece of material and keep the quality of construction at whatever level you want. You are also able to make more design changes during the course of construction since the house is being cut and assembled as you go. Some types of designs may not be available in a kit home and you will need to build the entire house yourself. Sometimes there is a certain feel you want to instill in the house, one of handcraftsmanship or maybe high tech, that kits do not offer. Finally, many people just want the satisfaction of knowing they built it themselves.

Manufactured Housing

Until several years ago, I, like many other people, had a certain attitude about manufactured houses. Since I began as a custom builder, I felt manufactured houses limited people's creativity and design and often produced an inferior product. In the last few years, I have made a 180° turn. Most all of the existing manufactured housing companies, especially those that have been in business for some time, offer high-quality products and the design options are limitless. On top of that, it is often a less expensive way to create the same quality house you would build for yourself. Here are several of the advantages to manufactured housing.

One of the greatest advantages to the owner builder is that you are linking yourself with a support network, the manufacturer and your local dealer, who will assist you through all stages of construction—from getting your permit, to getting finances, to getting the house built, etc. This support system can be of great value to a novice builder.

Your construction time is greatly reduced since a lot of the work is done at the factory. And if you are working on a limited time budget, this can be very helpful to you. A lot of the work is done by the manufacturer; the amount of knowledge and information and work you will have to do is limited. Often, a manufactured home can go up in a third of the time of a job-site-built home. This reduced time can be translated into dollar savings since your construction loan, which is often a few percentage points over your mortgage loan, will be reduced. If you own another home and are paying a mortgage while you build, you will be able to move into your new home sooner, eliminating the cost of two mortgages.

Manufacturers are able to buy in large lots and can get superior-quality materials at a lower price. This saving is often passed on to the owner. Since factory labor is a lot less expensive than hired carpenters, the work done on the house is less costly. This is passed on to the buyer.

For the most part, a house using high-quality materials can often be had more cheaply from a manufacturer than one built by yourself at the job site.

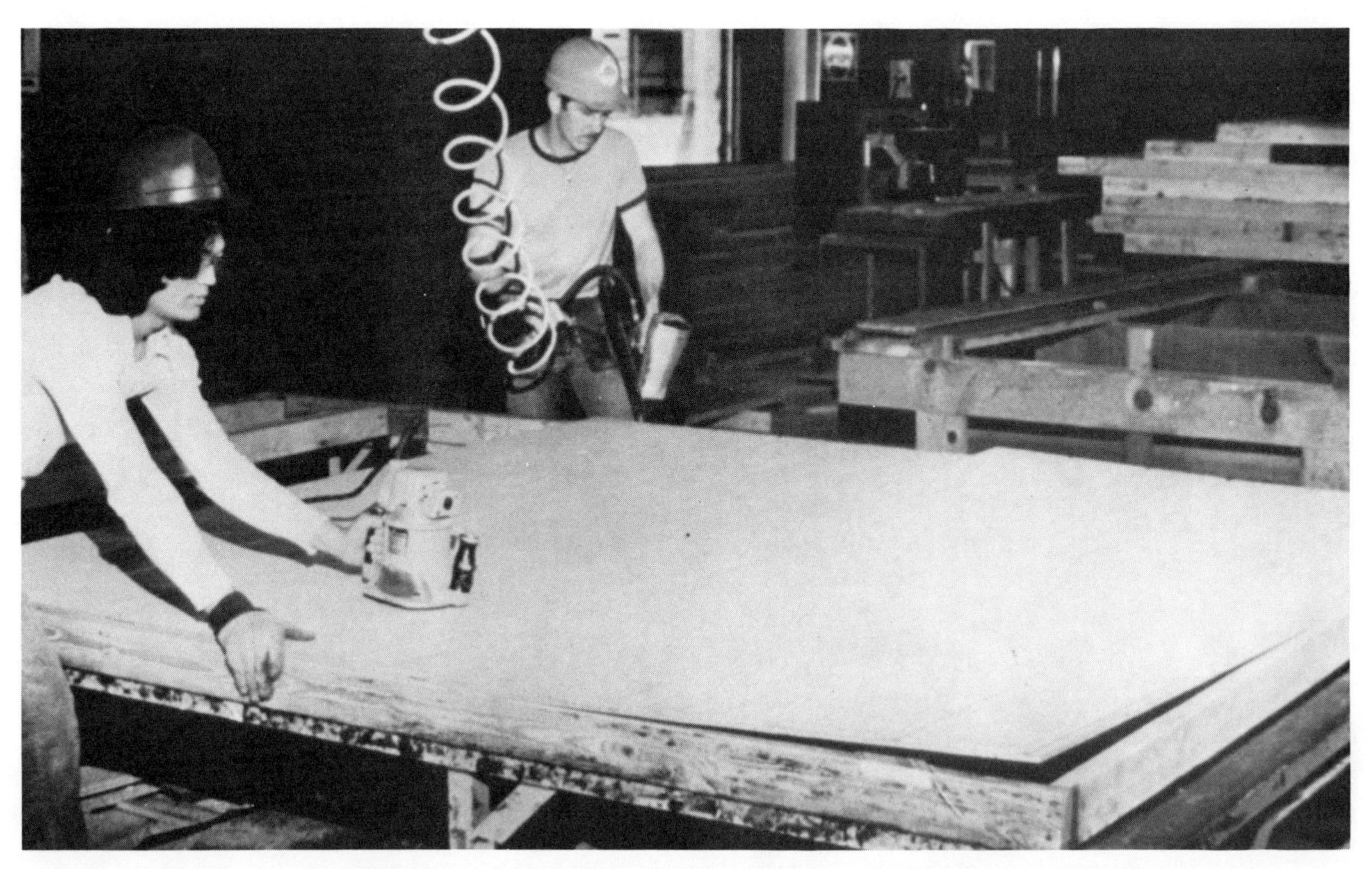

Photo 1. A modular wall section of a manufactured stud house being assembled in the factory with pneumatic tools.

Photo 2. A panelized wall section of a manufactured stud house being lifted into place by a crane. The entire house might be completed in a week or two.

Photo 3. A modular house being delivered to the site in several large sections.

There is one other added advantage. If you are building a home in a remote area where labor is expensive and unavailable, you can solve many of your problems by using a manufactured house.

A few problems with manufactured housing are that you may not always be able to get the design you want from every manufacturer, and occasionally there may be code problems in your local area with some manufactured housing. However, most of the larger manufactured housing companies have ICBO (International Congress of Building Officials) numbers, which will assist you in getting whatever permits you need to build the house in their area.

There are many different types of manufactured housing. Almost every style of home, including domes, logs, laminated timbers, stud, pole, etc., comes in a manufactured housing option. Some manufacturers simply provide the pieces for the weather-type frame of the home. Sometimes the entire interior finishing materials are supplied as well. In some panelized packages, the entire wall section, including the interior finishing boards, the exterior walls, the plumbing and electrical fixtures, the frame, and insulation, are put together in panelized sections, shipped to the site and assembled (usually with the use of a crane). Sometimes the pieces are numbered, and you erect the entire thing like a giant erector set. Other modular homes have the entire house in two or three large sections with the carpets down, the appliances in, and the plumbing and electrical fixtures in, that they bring to the site and bolt together.

I highly recommend that you explore the manufactured housing alternative for any style house you are planning to build. Here are a few tips to consider before choosing whether to use this alternative and, if so, which manufacturer to choose.

1. Be sure the manufacturer's package will comply with the local codes.
2. Investigate the company. What is their credit rating? How long have they been around? Do they have a record with the Better Business Bureau? Have they ever been taken in front of a consumer affairs board? Have they ever been bankrupt?
3. Ask for references of other customers who have recently built and lived in one of their houses.
4. Ask to see all of the instructional materials and educational pamphlets.
5. Be clear as to what the package includes and what it does not include. Make sure you have worked this out in detail and know exactly what you are getting for your money and what you will still have to go out and purchase or subcontract. Often a lot of problems lie in this area. It is hard to compare one manufactured housing company to another unless you know they are both offering the same types of materials and services.
6. Be sure that you will be able to re-sell this style of home in your area should you ever move.

7. Investigate whether or not you will be able to make design changes or design the house exactly as you want it.
8. Find out what other assistance the company will provide. Will they assist you with financing? Will they assist you with permits, with site choosing, with site improvement, with construction, with finishing?

Standard Stud Construction

Most of the houses built in the U.S. today are built using a standard stud construction. There was a time when almost all houses were either post and beam or log, but stud construction caught on with such a fury that, until recently, all others were almost forgotten. This did not come about because stud construction was, across the board, a better way to build, but it was quicker and easier, much as white bread replaced whole grain. The old construction method of post and beam makes use of large structural framing members, and therefore fewer timbers are needed, but construction usually required the skill of a highly trained joiner since a small number of joints carried the entire weight of the building. Because of this, each joint had to be exactly fitted. At the turn of the century there was a switch to stud construction; the work was quicker and took less skill. As cities grew, buildings needed to go higher in order to make better use of the expensive land. With the old post and beam method you would need longer and longer timbers in order to go higher. With stud construction, you simply stack one story on top of another, never needing a wall member over 8 feet long.

To build with stud construction, you first build a foundation, then on this you build a floor frame. The floor frame is built with 2-inch-thick construction-grade lumber (2×6, 2×8, 2×10 or 2×12) called floor joists. A floor joist is one of a series of parallel framing members that run underneath and support the floor. These are usually placed on 16- or 24-inch centers. After the joists are in place, you apply the subfloor (usually plywood) and then use this platform on which to build the rest of the house (hence the name platform building, as it is still occasionally referred to). The walls are built with studs, which are 2×4s or 2×6s that are about 8 feet long and again are placed on 16- and 24-inch centers. Where the structural integrity of the wall frame is violated by an opening for a door or window, a larger header spans across the top of the opening to carry the load to either side of the opening. Since the walls are carrying the roof load, these headers are essential. The roof frame is made up of roof rafters, which are also two-inch-thick stock that support the roof sheathing and roofing. These rafters rest on top of the walls, and the walls carry the load into the floor and onto the foundation and into the ground. If there is a second or third floor, you simply build your first-floor walls, place your second floor on top of them and then build your second-floor walls. The third floor is

Photo 4. The floor joist system of a stud house. In this case, the joists are cantilevered as a design feature.

Photo 5. A stud frame house under construction. Note that the platform is built and walls are going up.

Photo 6. A stud house under construction.

Photo 7. A stud house using some post and beam in the front section. Note that large framing members are used for the wall, but the rafters, since they will not be exposed to the interior, are 2×10s.

Photo 8. A stud house with some post and beam construction.

Photo 9. A post and beam house built by students at Heartwood Owner Builder School in Washington, Massachusetts. Note that the horizontal beams and vertical posts shown constitute the entire structural framing system.

added in the same way. So the roofing is supported by the rafters, which are supported by the walls, which are supported by the floor frame, which is supported by the foundation, which is supported finally by the ground.

There are many advantages to this type of building, especially for the novice builder. Everything in the construction world, at least as far as residential construction is concerned, is set up around this type of construction method. The materials and information are readily available and it will meet all local codes. In addition, since many different small pieces of wood make up the frame of the house, the load is spread over many joints and many pieces of wood. No one joint is essential and therefore a ¼-inch margin of error is allowable at the framing stages, taking the burden off the builder to make exact cuts. Finally, many of the building materials are provided in 4-×-8-foot sections: plywood, paneling, sheetrock, etc. These accommodate the 16- or 24-inch centers of the framing members. Heaters, ductwork, medicine cabinets, etc., are made to fit between the framing members.

Post and Beam Construction

Post and beam construction, which is sometimes called plank and beam construction, was used in the 18th and 19th centuries and was the dominant type of construction before stud construction was introduced in the United States.

This type of construction involves large timbers, usually 4 to 12 inches wide, that are placed both vertically and horizontally. The vertical timbers are called posts and these support horizontal timbers called beams. The load of the roof is carried onto the horizontal beams, then over to the vertical posts and onto the ground. The walls themselves do not carry the loads; they are used only to hang the siding, the doors and windows. The posts and beams carry the entire load. Therefore, fewer but larger pieces of framing materials are used in the standard stud construction. Usually in this type of construction, the foundation and floor system are built as in conventional stud construction and then the posts and beams are erected on the floor platform. Many times post

Photo 10. A metal bracket used to hold the post and beams together in earthquake areas.

Photo 11. Large timbers are bolted to poles in this pole house. Note how the house is resting on these timbers.

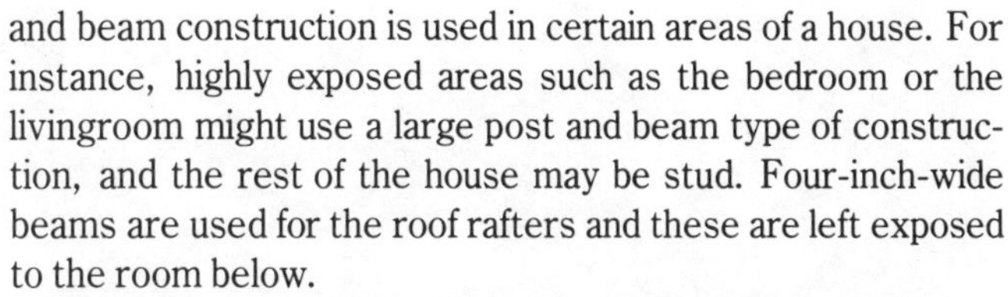

and beam construction is used in certain areas of a house. For instance, highly exposed areas such as the bedroom or the livingroom might use a large post and beam type of construction, and the rest of the house may be stud. Four-inch-wide beams are used for the roof rafters and these are left exposed to the room below.

The greatest advantage of post and beam construction is the sense of solidity, character and integrity that the large exposed timbers lend to the structure itself. A home done with a stud construction, without the large exposed framing members, will feel quite different from the same house built with large exposed framing members. Where posts and beams, rather than the walls, carry the load of the roof, you can also have larger and more frequent openings for doors and windows, and large interior spaces. Post and beam houses can be designed to fit almost any style. Usually 4 to 5 people and perhaps even a crane may be needed during the frame-construction period.

The disadvantages in post and beam construction are few. In some areas, engineering may be required, but this is usually not too costly. It can also be difficult to find high-quality larger timbers. Since there are fewer framing members and each one carries a greater load, the cuts and your general work must be more exacting than in stud construction, where many small framing members carry the entire load.

In earthquake areas, as well as in hurricane and tornado areas, where the house stands a lot more stress and strain than in areas where these natural disasters do not occur, houses have to be better built. Because of this factor, post and beam homes often have to be connected with large metal brackets. This is especially true in California. In this case, your work does not have to be as exacting since the joints are not made of wood but of metal brackets, bolted to the wood and framing members. Yet even this can have a very distinctive appearance which should be considered. As in all other types of construction, post and beam packages are available from a number of different manufacturers.

Pole Construction

Pole construction involves vertical load-bearing poles that are embedded in the ground sometimes up to 6 or 8 feet deep. These poles continue up and support the floor frame, the second floor and the roof. The floors and the roof are attached by horizontal boards that are bolted and often notched into the vertical poles. This type of structure is becoming more and more popular. In the past, it has often been used for agricultural and farm use where the dirt floor between the poles was used for equipment and a place where livestock could move around within the building. Given some of its unique advantages, pole construction is now having somewhat of a renaissance.

Perhaps the greatest advantage of this type of construction is its practical use on a sloping lot. Often on sloping lots foundations can be very expensive; as the lot slopes, the

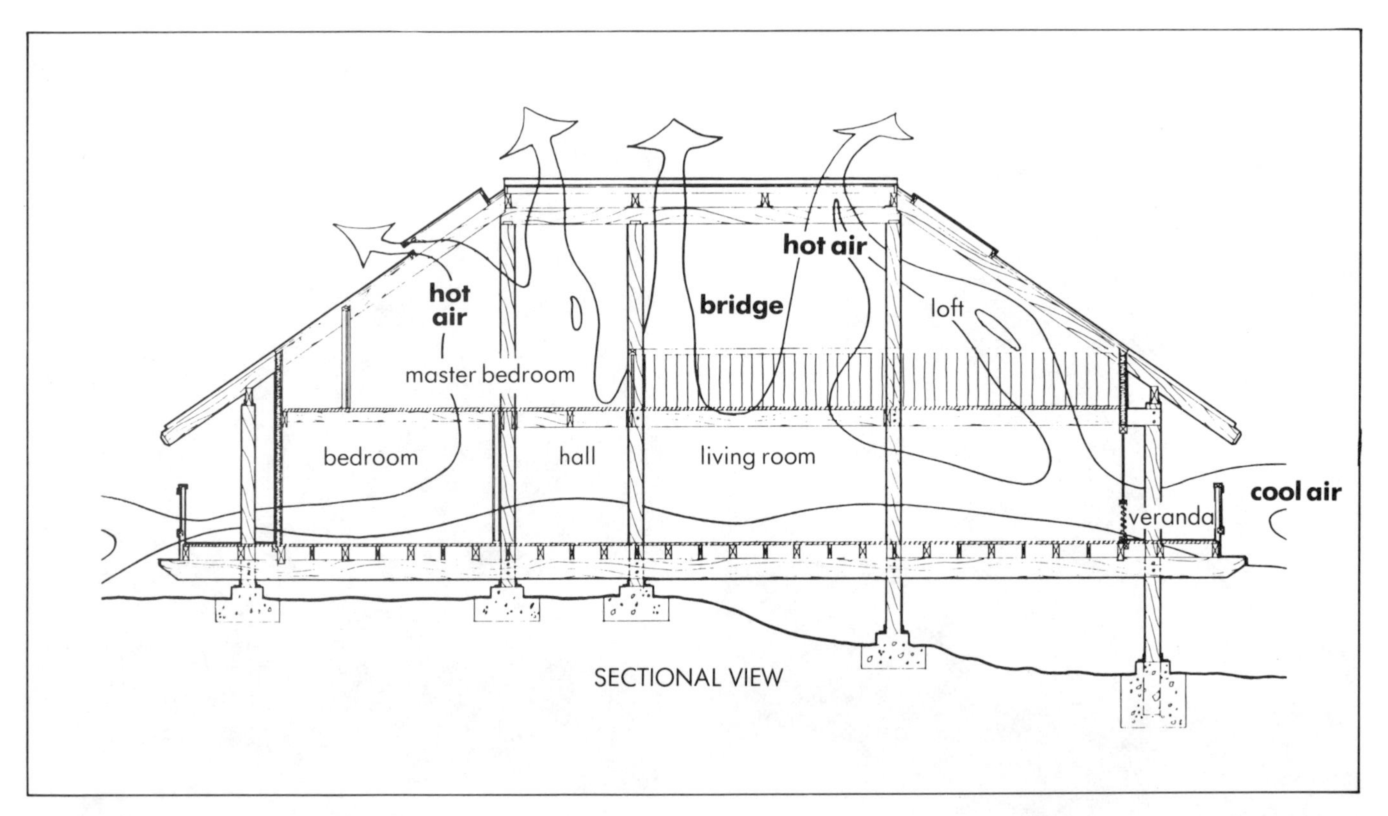

Figure 1. Since pole houses do not require as many interior bearing walls as standard construction, they are very well suited for natural ventilation in summer and transfer heat easily between rooms in passive solar homes. (Diagram courtesy of Pole House Kits of California, Irvine, California.)

Photo 12. A pole house in progress. Note how the walls are framed in typical stud fashion.

Photo 13. Poles for this pole house are kept out of the ground by fastening them to concrete piers. Note the small redwood sill block.

Photo 14. A pole house under construction. Note how few framing members are actually needed for the structural components of the house. The large poles serve as foundations and support the floors and roofs. Standard-size framing members will be added between the poles to support windows and doors and attach interior and exterior skins. The roof was added to protect the home early in construction. (Photo courtesy of Pole House Kits of California, Irvine, California.)

foundation must become higher and higher to provide a level plane. With pole construction, however, longer poles are used at the lower end of the lot and shorter poles at the higher end to make adjustments for the foundation. This can save you many thousands of dollars in construction costs. A pole house also performs very well in hurricane, tornado and earthquake areas since the house is tied well into the ground with the long poles acting as the foundation.

Like post and beam construction, it uses some large framing members to give the house a sense of strength and solidity, but the exactness of post and beam work is not required with the round poles. The house can go up rather quickly and the roof can be put on before the walls are completed, thereby providing a protective work area as you finish the inside of the house.

Often it will take several people and perhaps even a crane to erect the large poles, but after this the work goes rather quickly. It is best to use some type of pressure-treated poles so the wood will not rot in the ground, or to build a system where the poles sit on top of a concrete pier foundation and are bolted to this foundation. This will also keep the wood out of contact with the ground.

Perhaps the greatest advantage of the pole house, aside from the strength and solidity of the large poles, is the openness it can allow. Load-bearing walls are not needed, as the poles carry the weight, and both large interior and exterior openings are possible. Also, large sweeping verandas are very simple to implement in your house design.

Though in the past most people have chosen to build their own pole house, nowadays there are a few kit companies available. If you are planning to build a pole house, it may be a wise decision to investigate one of these kit homes to see if there is a labor and money savings.

Photo 15. Pole houses are especially well suited for sloping sites, where the longer poles are used on the lower part of the slope. The savings over traditional foundation systems can be considerable. (Photo courtesy of Pole House Kits of California, Irvine, California.)

Photo 16. A finished pole house. Once again notice its application on steep slopes. (Photo courtesy of Pole House Kits of California, Irvine, California.)

Photo 17. Pole construction used for a commercial building. Note how well the poles work on a sloping site.

Photo 18. Pole houses are conducive to implementing wide-covered verandas as shown here. Note stringer attached to pole at upper right. (Photo courtesy of Pole House Kits of California, Irvine, California.)

Photo 2-19. Pole houses allow for open spaces and cathedral ceilings. The poles carry the load and can allow for large open interior areas without interior bearing walls. (Photo courtesy of Pole House Kits of California, Irvine, California.)

Log Homes

Log homes were one of the first types of construction to come to the United States. All of us were familiar with the log homes built by the pioneers and our ancestors when they founded this country. Today there are still over 80,000 log homes built each year in the United States.

Log homes have a feel different from almost any other type of structure. Their massiveness and strength is apparent as soon as you walk in them or view them from the outside. A house built with a stud construction will feel entirely different from a house of the same design built from logs.

Log kit homes come in several different types. One type includes the walls only. This is usually made up of precut logs and does not include a roof system, doors, or windows, but simply the logs for the walls. Another type of kit includes the logs for the walls, the roof system, the doors, and windows, providing therefore a weather-tight shell. A third system includes the weather-tight shell, the floors, the interior walls, counters and stairs. And finally, the last system includes everything you need for the completed house.

Log homes are usually rather simple to erect and require less skill and exactness of construction than many other types of homes. It is really just like putting together a large lincoln-log set that we used when we were children. One great advantage of erecting a log home is that as the logs are erected, you are not only framing the house, but you are also insulating the house and finishing off the interior and exterior walls. As a bonus, log homes require little maintenance, usually just some wood preservative every few years.

Photo 20. Pole construction can also be combined with the beauty and strength of post and beam construction, as shown here. (Photo courtesy of Pole House Kits of California, Irvine, California.)

Photo 21. A hand hewn log building. This type of structure can be inexpensive to build and of greater strength and quality than typical stud construction.

Photo 22. A manufactured log house kit. Beautiful and solid.

Photo 23. The logs from a kit home stacked and ready for assembly.

Photo 24. Round log home built by owner from logs cut at the site.

There are many, many different log-home manufacturers and the design variations have multiplied considerably in the last few years. Log homes, being easy to assemble, usually require no more than 2 or 3 people and very novice skills.

Many people are concerned about the energy efficiency of a log home. Wood has an R factor of about 1.5 per inch, whereas good insulation can have an R factor of anywhere from 4 to 6.5. (R=Resistance to heat loss; the higher the R number, the greater the insulating property of the material.) Therefore, the R factor of a typical log wall is often no more than 8 to 10 while that of a stud home can be 11 to 14. This has discouraged some people from building a log home. You may want to consider building a laminated timber home, mentioned in the next section, which gives you the advantage of a solid-wall wooden home and at the same time provides the increased R factor of 23 or even higher. However, many people who live in log homes say that they are very comfortable in winter and summer and that the R factor of a true log home is not the only thing to be considered. The cellular nature of the solid wood acts as a much greater natural insulation by trapping air in the cells, making a log home more comfortable than the R factors would indicate.

Before choosing to build a standard log home, be sure that it fits all local codes. Usually this will not be a problem since log homes can exist anywhere. It should blend-in in your area. If you are building in the urban or suburban area, often times your design style will not fit in with that of your neighbor's.

Try to choose a dealer who can transport the heavy logs to your site without a tremendous increase in transportation costs. And, above all, be sure that you feel comfortable in the solidity and strength of a log home and that the tremendous amounts of exposed wood will go with your decorating scheme.

Laminated Timber Construction

In the past few years, a new manufacturing process has been developed which gives the solidity, quality and strength of a solid log home or solid wood home, and at the same time gives a greater insulating factor than typical stud construction. The process involves laminating cedar to both sides of styrofoam (or other types of insulating foam) core. Usually 2 inches of cedar are used on each side of 2-inch-thick foam core. The R factor of this wall can be anywhere from 19 to 23 as compared with an R factor of 13 for a typical stud house.

These houses have the advantage over log homes in the sense that they provide a greater insulating property while still offering all the advantages of solid wood. They have an advantage over a hollow-wall stud home in that they are solid and add a much greater sense of stoutness and strength to the house and the hollow core of a stud home.

Usually they cost little or no more than a similar house of the same quality built in a stud or log construction. Another advantage of these houses is that, as you close in the house, you are finishing the exterior and interior walls, framing the home, and insulating it all in one process, much as you do with the stud home.

Photo 25. Mock-up of intersecting laminated walls. Note that the insulating foam core, sandwiched by cedar planks, offers a solid wall with good insulation. (Photos courtesy of Pre-Cut International, Woodinville, Washington.)

Photo 26. Cutting the laminated logs in the factory from a customer's design. Each house is cut according to the customer's own plans and then marked as to its location in the wall. (Photo courtesy of Pre-Cut International, Woodinville, Washington.)

Photo 27. Electrical wires are run through the bottom few courses of a laminated timber home. (Photo courtesy of Pre-Cut International, Woodinville, Washington.)

Photo 2-28. A laminated insulated timber home under construction. Note how no trimmers or headers are needed around the openings. (Photo courtesy of Pre-Cut International, Woodinville, Washington.)

Photo 2-29. A laminated timber house under construction. Logs are cut at the factory from the owner's plans. (Photo courtesy of Pre-Cut International, Woodinville, Washington.)

Photo 2-30. The interior of a laminated timber home. Interior walls look like walls covered with tongue and groove cedar planks. Interior walls can also be covered with traditional wall coverings. (Photo courtesy of Pre-Cut International, Woodinville, Washington.)

Photo 2-31. Finished solar home built from laminated insulated timbers. The exterior appears as a home sided with tongue and groove cedar boards. (Photo courtesy of Pre-Cut International, Woodinville, Washington.)

The sound transmission control, which represents the ability of the house to reduce the sound from within and without, is much greater because of the solidness of the home and the foam core. Its shear strength and wind-load factor is superior to that of the stud home. The wood also provides natural beauty to the home, and since wood breathes, it acts as a humidifier and dehumidifier. It is easier to maintain, requiring only a wood protection in the beginning and a light coating of a wood preservative every few years after that.

Energy wise, it saves tremendous amounts of energy due to the increased R factor and the continuous insulating membrane provided by the solid home. Where a typical stud house uses 5,000 to 7,000 board feet of common-grade wet lumber, a laminated timber home uses 18,000 to 22,000 board feet of kiln-dried timber. This wood has minimal shrinkage and more strength.

The house can be designed according to your design plans, or it can be chosen from the manufacturers. You can design whatever style house you would like, anywhere from a southern mansion to a modern contemporary home, and it can be cut to fit this style. Should you choose to paint the outside of the home, it would only appear as if the house was sided with 1-×-8 siding boards put on horizontally. On the inside, the interior partition walls can be made out of sheetrock and even the interior surfaces of the exterior walls, which are usually made out of cedar, can be covered with sheetrock in order to make the house have a more traditional look. Both the exterior walls and interior walls can be left with the cedar exposed, giving a more rustic appearance.

Most of these houses are approved by the International Congress of Building Officials and will have no problems passing your local code.

In many ways, this is a superior building style. It gives you the solidity and strength of a solid timber home without the drawback of a log home, the lack of an insulating property. At the same time it allows you a tremendous amount of design flexibility and can appear to be a typical house built like any other house. It costs you the same price, but ends up to be a much higher-quality home.

Photo 2-32. Completed laminated log home.

Photo 33. A laminated timber home in traditional style. Exterior and interior look like T&G cedar siding. (Photo courtesy of Pre-Cut International, Woodinville, Washington.)

Photo 34. A dome shell under construction. A shell can be erected from a kit in a day or two. The triangular pieces are bolted together to form a pentagon which is then bolted in place to other assembled pentagons. (Photo courtesy of Cathedralite Domes, Medford, Oregon.)

Domes

Domes are free-standing structures which need no load-bearing interior walls to carry the weight of the roof or the structure itself. They were developed in 1956 by Buckminster Fuller and Al Miller.

Domes have several advantages. For one thing, the structure encloses the maximum amount of living space with the least amount of exterior surface area for its volume. Also, domes are now accepted by most code offices throughout the United States and dome manufacturers usually have an ICBO (International Congress of Building Officials) number that allows them to pass all local codes. Presently there are over 200,000 domes in use world-wide and it is predicted that over one million domes will be erected during the 1980s.

There are two different types of dome kits: the panelized and the hub and strut systems. The panelized dome kit is made up of triangular panels, usually made of plywood and 2×4s or 2×6s, with struts on all three sides and exterior skins already in place. These plywood triangles are usually one of two sizes and are joined together in pentangles which are then lifted into place. Usually 3 to 5 people are needed and scaffolding is erected on the dome floor to lift these pentangles into place. Often a riser wall, which is a vertical wall 3-to-5-feet tall, is built and the dome is erected on top of it. This gives the house some vertical spaces and also allows for a larger home. Openings for doors and windows are placed in different areas around the dome and skylights can be added by simply leaving out one of the triangles in the dome surface itself.

Another system, though not as commonly used, is the hub and strut system, developed in Arizona in the mid-1960s. In this type of system the frame of the dome or the skeleton is first erected similarly to that of a stud house. The frame is usually made up of 2×4s or 2×6s bolted together with metal hubs or plywood hubs. The external skin, which is usually plywood, is then applied just as the sheathing for a stud house would be applied over its frame.

Both kits usually contain only the shell with the interiors, the foundations and the finishing work being supplied by the owner.

Dome homes are some of the most energy-efficient homes

Photo 35. Domes can be used for commercial and industrial application as well as residential. (Photo courtesy of Cathedralite Domes, Medford, Oregon.)

Photo 36. A completed dome at the seashore. (Photo courtesy of Cathedralite Domes, Medford, Oregon.)

Photo 37. A completed dome built above a full-size first story with a slab foundation. (Photo courtesy of Cathedralite Domes, Medford, Oregon.)

Photo 38. The interior of a residential dome. Note how the shell can span the entire house without the use of interior bearing walls. (Photo courtesy of Cathedralite Domes, Medford, Oregon.)

Photo 39. The upstairs interior of a dome. Again note vaulted ceiling's interior partition walls. (Photo courtesy of Cathedralite Domes, Medford, Oregon.)

Photo 2-40. The floor system of a hexagon is a series of radiating 4-inch-thick floor beams. (Photo courtesy of Honeycomb Systems, Inc., San Juan Bautista, California.)

that you can build. Since the maximum amount of space is enclosed by the minimum amount of surface, the heat loss is cut to a minimum. Dome manufacturers claim that they will out-perform any other type of structure with a comparable amount of insulation. Domes provide natural air circulation, they have good acoustics, and they require less maintenance, since there are no corners for moisture or dry rot to build up in. Unlike many types of homes, no interior walls are needed to support the roof. Indeed, the entire dome can be one open room. Because of this you have a much greater design flexibility. Often the shell can be put up in a day or two and allow you to work inside with protection from the weather. Domes can also withstand a great amount of wind and snow and, because of their stable shape, they are very earthquake resistant.

The structure of a dome is entirely that of one large roof rather than walls and roofs as in many types of construction. Roofing, therefore, must be done properly to avoid leaks. Almost all manufacturers have developed good roofing techniques and this is no longer a problem, but be sure you do the roof properly. Since there are few vertical walls but many sloping walls and angles, decorating can be a challenge in a dome home.

One problem with dome homes might be that financing may be difficult to obtain. However, more and more banks are getting used to seeing this type of construction, and if it is a problem with your local bank, with the help of your manufacturing company and perhaps some other local dome owners, you should be able to convince them to finance the home.

Before deciding to build a dome home, go and visit several different domes. From my experience in dealing with owner builders, I have found that most people who build a dome know almost instantly upon going in their first dome that this is the type of housing they want to build. They feel more comfortable with the angles and the sloping walls than they do with the true vertical and horizontal planes and the 90° angles that most other houses offer. In making this important decision, ask yourself these questions: How does the dome feel to you? Does it make you feel at peace? Do you feel more at rest in a dome, as opposed to a conventional type of structure?

Photo 41. The completed framing of a hexagonal home. (Photo courtesy of Honeycomb Systems, Inc., San Juan Bautista, California.)

Hexagonal Homes

Another type of construction that is available in kits in the United States is hexagonal construction. As the name implies, this type of construction involves 6-sided buildings. This structure creates a rather unique living space with almost no right angles anywhere in the structure. Because of this and the fact that the structure needs no interior supporting walls, your design parameters and the feel of the house can be quite different from other types of construction.

The foundation system is built in a hexagonal shape. The floor system is built on top of this using beams that radiate in a hexagonal shape. A subfloor is applied and posts are erected around the exterior of the house. The posts will support the large roof beams. At the center of the home these roof beams are supported either by more vertical posts or by a "truss ring"—no vertical posts are needed in the center of the home. Walls constructed between these posts simply serve the purpose of attaching the siding, doors, windows, insulation and interior wall covering. Like the dome, the hexagonal structure uses the strength of the triangle to give it stability and flexibility. It is a series of interactive triangles from which it derives great strength.

Hexagonal homes are available from several kit manufacturers or can be built by the owner from scratch. Since the pieces are all cut from patterns and the design is a simple one it is easy for the novice builder to build. Materials are easily available and it will usually meet all local codes. Since the weight of the roof is carried by the wall posts and not the walls themselves, windows and doors can be easily added and deleted. The buildings can be built up to three stories high or can be clustered together in a series of small hexagons. This allows you to add space as it is needed or can be afforded. Because of the vaulted ceilings and lack of right angles, the interior spaces can often feel larger than they really are. In many ways these homes reflect a certain change in lifestyle, one that is individual and yet strives for unity and order.

Photo 42. Hexagonal buildings have a good view from all sides. (Photo courtesy of Honeycomb Systems, Inc., San Juan Bautista, California.)

Photo 43. Hexagonal buildings can be built in clusters. This makes it easy to add on as the family grows or more funds become available. (Photo courtesy of Honeycomb Systems, Inc., San Juan Bautista, California.)

Photo 44. The interior of a hexagonal home. The large ceiling beams rest on the structural thrust ring, a feature which provides integral strength and flexibility to the hexagonal home. (Photo courtesy of Honeycomb Systems, Inc., San Juan Bautista, California.)

Photo 45. The interior of a hexagonal structure. Note the open spaces and the lack of right angles. (Photo courtesy of Honeycomb Systems, Inc., San Juan Bautista, California.)

Included in this chapter are pictures (Photos 46–51) of various styles of homes popular in different parts of the country.

Photo 46. Natural stone houses are rare, but permanent. These houses are labor-intensive, but beautiful.

Photo 47. Partial cut stone and post and beam structure.

Photo 48. Brick structures are common in some areas. Often stud frame houses are covered with a brick veneer.

Photo 49. Concrete block buildings are common in many areas. This one is in Florida where termites would pose a threat to wood frames.

Photo 50. A-frames are a quick, easy and inexpensive type of construction. I built this one in North Carolina in 2 months. It is 1200 square feet.

Photo 51. A house I built from a large wine barrel. Six inches of wine was left in it to create a "wine cellar."

14/Tools

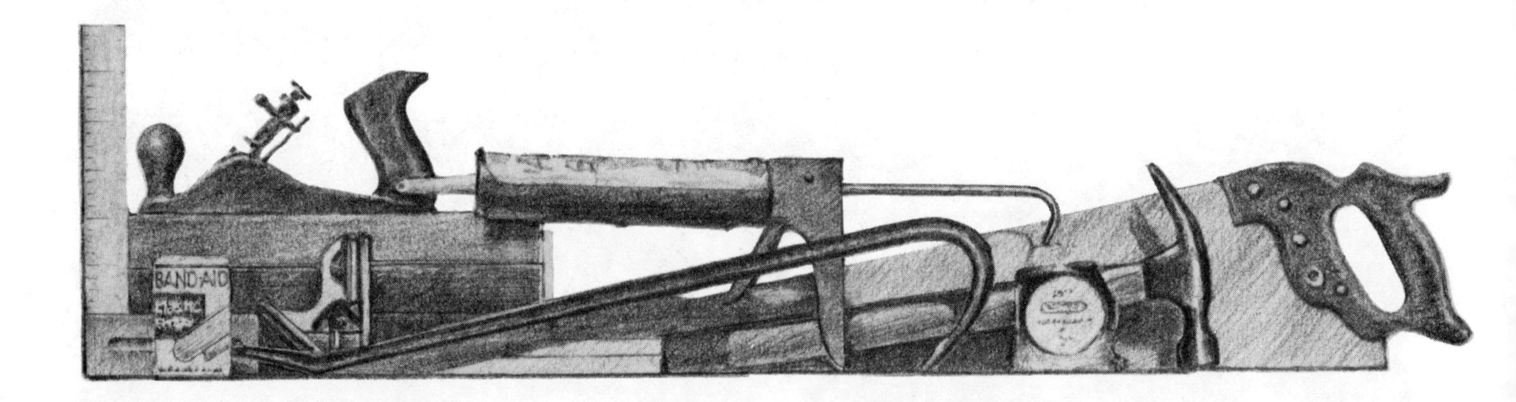

The following section on tools was written by Mel Berry. Mel has taught people to remodel their homes for many years and is a licensed cabinet maker.

THE FIRST CONSIDERATION, before starting to build portions of a house, should be to get good quality tools and understand how to use them. Frequently those who have not done a large construction project have gathered a collection of "used" tools made up of other people's cast-offs, slightly broken and usually dull. They have been satisfactory for putting a door back on its hinges, cutting firewood, replacing a stair tread or a piece of decking, but now the greater task of many hours in housebuilding demands a careful evaluation of tools available and their condition. The most highly skilled craftsperson would not be able to do satisfactory work in a reasonable period of time if his tool collection was not as carefully developed as his work habits and practices. The pleasures of building and creating as a craftsperson will change to frustration if dull screwdrivers tear out screw slots; dull chisels cause errors and even injury because of excessive pressure and force; dull hand saws leave one's arm aching and hand blistered; and dull power cutting tools smoke, burn and kick back.

Buying Tools

A builder should get the best tools he can afford. Choosing well-known brands is probably safer for a novice. In comparing both power and hand tools, consider balance, weight, comfort and durability. An electric drill with too small a handle feels "top heavy." A 32-ounce framing hammer for a large person will drive nails quickly, but would be too much for a smaller person. A chisel with an all-plastic handle will not take as much pounding as a chisel with the steel of the blade extending all the way through the handle. In most cases the better tools will cost more, but it is a necessary expense and actually small, considering the size of the housebuilding project. Also, better tools frequently carry manufacturer's warranties and even lifetime guarantees. Once sharpened, good cutting tools keep a sharp edge. (Photo 3-1.)

Sharpening Tools

Most carpenters today have their saws sharpened by professional sharpening companies, but they sharpen their own plane irons, knives, screwdrivers and chisels. Regrinding plane irons, knives, and chisels is not usually needed unless the tool has hit a hard object during use and has chipped. If regrinding is necessary, it can be done by some saw sharpening shops or the carpenter can re-do the shape if he has his own bench grinder. For routine sharpening of tools, a craftsperson needs metal files, a combination bench stone and honing oil. Metal files are used for reshaping screwdrivers, hatchets, shovels, chainsaws and other tools. For some tools, filing is satisfactory sharpening. Other tools need honing on a bench stone.

A combination bench stone has a coarse side and a fine side. Use the same procedure for sharpening chisels and plane irons. With the bevel at the angle of the original grinding, move the tool in a figure 8 pattern over the surface of the stone until the edge is improved, usually about 30 to 40 strokes. Honing oil applied to the surface of the stone keeps the metal from filling the pores of the stone. Then turn the stone over and continue honing, using the fine side of the stone. The back or flat side of the tool is rarely ground or stoned, except for a few light strokes occasionally to remove

Photo 1. Avoid cheap tools. They can be dangerous when they fail.

the metal that curls over during sharpening, creating a burr. The figure 8 pattern on the stone is not essential, but helps to keep the oil distributed. If the stone gets "dirty," or clogged with sawdust, it must be cleaned with a solvent, such as kerosene. Other stones that can be used for fine honing include the natural Arkansas stones and Japanese water stones. After sharpening, a tool should be stored in a protective covering to keep its fine edge.

Basic Tools

These basic tools should be collected before attempting to build. They will all be needed frequently.

Combination stone
Metal files
20- or 25-foot tape
Plumb bob
Chalkline
Pocket knife
Utility knife
Carpenter's pencil
Carpenter's square
Combination square
2-foot level
Crosscut saw
Rip saw
Coping saw
Hacksaw
Wood chisels
Cold chisels
Diagonal cutting pliers
Nail sets
Hammer
Screwdrivers
Phillips screwdrivers
Pliers
Adjustable wrench
Block plane
4-in-1 file
Flat bar
Nail claw
Twist bits
Spade bits
Circular saw
Sabre saw
⅜ variable speed reversing belt sander
Finishing sander
Heavy duty extension cord

Secondary Tools

Secondary tools have less frequent usage than the basic tools, but most of them should eventually be added to a builder's toolchest. Until then, they will have to be borrowed or rented when they are needed.

Bench grinder
50- or 100-foot tape
Builder's level or transit
Sledge hammer
Bit brace
Hand drill
Power auger bits
Forstner bits
6-foot level
Framing hammer
Assorted files
Ripping bar
Countersink
Router and bits
½-inch right angle drill
Reciprocating saw
Power miter saw
Power plane
Com-a-long
Rabbet plane
Nail shooter
Aviation snips

Specialized Tools

Specialized tools are infrequently needed tools, or an expensive purchase that can only be justified with a great deal of use on a single job or repeated professional usage.

Laser level
Foundation stake puller
Hardwood flooring nailer
Shingle ripper
Battery powered drill
Pneumatic nailer and equipment
Pneumatic stapler and equipment
Radial arm saw
Table saw
Tool specialties from various trades

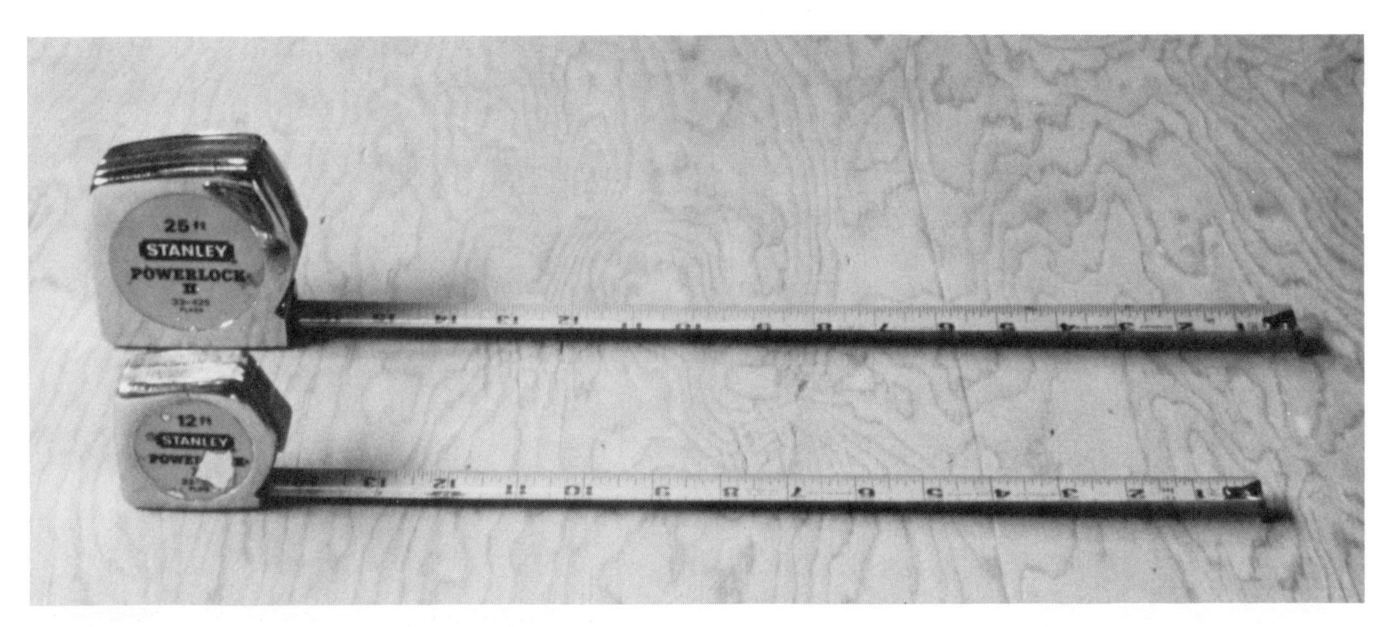

Photo 2. Metal tapes are an essential. I recommend a 20- or 25-foot tape with a 1-inch blade.

Photo 3. Here is an example of using a plumb bob to see if a post is plumb. Note that a block is used at the top and bottom of the string to see if it is an equidistance from the post. Plumb bobs are also used to locate points directly below other points.

Layout Tools

STEEL TAPES (Photo 2) have largely replaced the folding wood rules for most people. The usual lengths are 20 or 25 feet for general measurement, and 50 or 100 feet for larger layout jobs. The most desirable width for rough carpentry is 1 inch, since it allows the worker to extend the tape further before it "buckles."

A PLUMB BOB (Photo 3) is necessary to determine a reference of straight up and down or right angles to the earth. While any heavy object suspended on the end of a string will determine this reference by gravity, usually a craftsperson will purchase a machined plumb bob for greater accuracy. The heavier a plumb bob is, the less it will be affected by the wind while it is being used.

A UTILITY KNIFE (Photo 4) should be of high quality steel and fit comfortably in your pocket. Some craftspeople, however, prefer to have a larger knife and keep it in their tool pouch or on their belts. The choice is yours, of course, but a knife is necessary to cut cord, mark closely around hinges for mortises, sharpening pencils, etc. Most craftspeople will also have a utility knife for general use.

The CARPENTER'S PENCIL (Photo 5) is a flat pencil with large lead. It is easy to sharpen, draws broad lines that are easy to see and it won't roll away.

The T-BEVEL (Photo 6) or bevel square has an adjustable blade which can be set to any angle. It is usually used to transfer angles from the work spot to the cutting tools.

The COMBINATION SQUARE (Photos 7, 8) has several uses. Frequently it is used to mark a 90° line across a board to follow when cutting or when joining another board to it. It will also measure a 45° angle, check for square, and measure and transfer depths and distances.

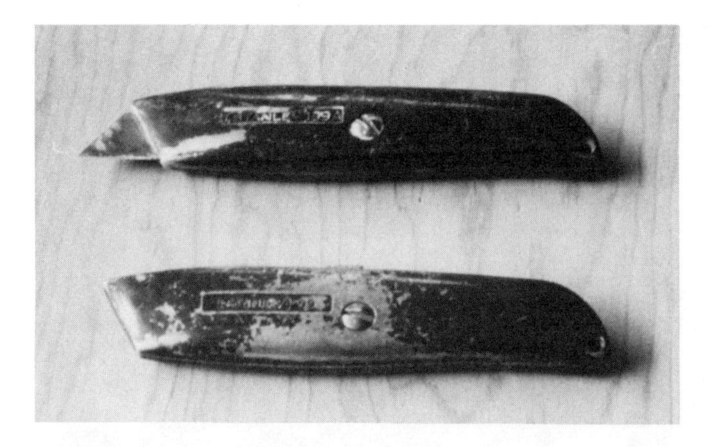

Photo 4. Utility knives are needed in many processes during construction, from cutting the layout strings to scoring the drywall. Be sure to get one that has a retractable blade; extra blades are stored in the body.

Photo 5. Carpenter's pencils are a must. They come in broad and narrow leads. Keep a supply on hand.

Photo 6. Using a sliding T-bevel to find an angle. This angle can then be transferred to the board you are cutting.

Photo 7. Using the combination square to mark a board at a 90° angle. This square can also be used to mark a 45° angle and to scribe the length of the board.

Photo 8. Using the combination square to mark a straight line down the length of the board. This method ensures that the line you are marking is an equidistance from the edge of the board even if the board is not straight.

Photo 9. Using a framing square to lay out a stair tread. Framing squares are used throughout the housebuilding process. Get a good one and be careful not to leave it lying around where it could get stepped on and bent. It is of no use once it is bent.

Photo 10. A chalkline is used to mark straight lines.

Photo 11. Using a 6-foot level to check the plumb of a post. Note the clips at the top and bottom of the level. These clips keep the level off the board which will have many irregularities and make it hard to determine true plumb. You will need a 2-foot and 4-foot level, and a 6-foot or 8-foot is advised.

Photo 12. A torpedo level or 1-foot level comes in handy. It's being used here to level a pier block.

Photo 13. Transits, or their cousins the builder's levels, are needed during the layout. You can rent these, but if you can afford one, they are very useful during many states of construction in checking for level and plumb.

The large CARPENTER'S SQUARE or FRAMING SQUARE (Photo 9) has many uses in addition to measuring squareness of work, as mentioned in other chapters of this book. Today most carpenters choose aluminum framing squares since they are lighter to handle. Some squares have one special scale measured in 12ths of an inch for reading scale drawings.

A CHALKLINE (Photo 10) is used to lay out straight lines for reference and assembly. It can be a length of line and a block of carpenter's chalk, or it can be a line inside a chalk box that applies powdered chalk as the string is drawn in and out.

Levelling Tools

A 2-FOOT LEVEL is used for spot checks of plumb and level. For more accurate measurements, a 6-FOOT LEVEL (Photo 11) is necessary. Some craftspeople extend the use of the 2-foot level by putting it on a longer straight edged board. A 1-FOOT LEVEL (Photo 12) can be used for levelling smaller surfaces.

A BUILDER'S LEVEL is not part of every carpenter's tool collection, but is essential for efficient layout of footings and foundation walls. The builder's level establishes a level line of sight; all points along that line of sight are the same as any other point.

The TRANSIT (Photo 13) has one more operating mode than the builder's level. It pivots up and down in a vertical plane and enables the operator to measure and transfer points in that plane. It can also be used to plumb work. These tools are expensive for one-time use and can be rented. (For an explanation on the use of a transit, see *Layout* chapter.)

A newer levelling device, the LASER LEVEL, projects a beam of coherent light onto surfaces for a reference indication. Laser levels, however, require electrical power, which may not be available in the initial layout of residential homes.

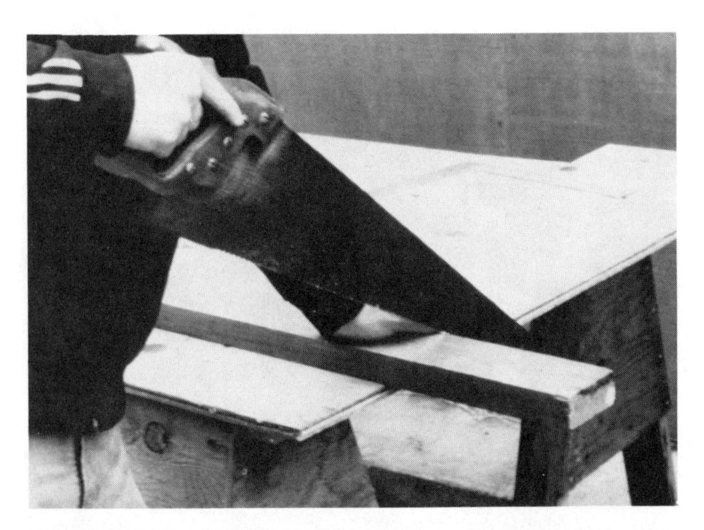

Photo 14. There are several different types of hand saws; their use depends on whether you are cutting across the grain (crosscut saw) or with the grain (rip saw). A finishing saw has teeth that are set closer together. You may want to buy a general combination saw that can cut in either direction.

Photo 15. Chisels are a must. I advise a ¼-inch, ½-inch, ¾-inch, 1-inch and a 1½-inch. Get the kind where the metal shaft extends all the way to the butt, as shown on the left. The metal shaft on the right is embedded in plastic and will not last as long.

Photo 16. Several kinds of hammers are needed during house construction, i.e., a framing hammer, finishing hammer, drywall hammer, etc. Use ones with fiberglass or wood handles.

Cutting Tools

SAWS (Photo 14). Today's craftsperson cuts mostly with electric tools, but a crosscut saw, with about 8 knife-shaped teeth per inch, is used when power is not available or convenient and when the round-cutting pattern of a circular saw is not acceptable. When cutting stair stringers for example, the circular blade will not cut completely into the corner of the layout (see section on stairs). A sharp crosscut saw may also outperform the circular saw if lengthy extension cord layout is necessary. A blade cover should be purchased or fabricated to prevent damage to the teeth. Finish saws are usually slightly shorter and may have up to 12 teeth per inch.

A RIPSAW cuts with the grain of the wood and has chisel-shaped teeth to keep the saw from following the grain. Nearly all ripping cuts today are done with a circular whenever possible.

A COPING SAW or hand jigsaw is used especially to shape moldings to fit. Its fine blade allows intricate cuts over short distances.

A HACKSAW is necessary to cut metal stock and can be used for wooden moldings.

WOOD CHISELS (Photo 15) from ¼ inch to 2 inches in width are necessary to remove stock for hinges, shaping openings and patterns. High quality steel is necessary to keep a sharp edge longer. Strong handles that will take hammer blows are essential for a carpenter since mallets are rarely available. A cold chisel is necessary to cut through nails and undesired metal pieces.

Assembling Tools

A HAMMER (Photo 16) is probably one of the first tools associated with carpenters. Many carpenters have several hammers for different purposes and enjoy their unique features.

FRAMING HAMMERS are available in weights from 16 to 32 ounces with ripping and curved claws. Although few people find the heaviest weights comfortable, choose the heaviest hammer that can be used without tiring. The straight or ripping claw is useful for driving between pieces and as a lever. The curved claw gives more nail-pulling leverage. Framing hammers are also available with waffled faces to prevent slipping from the nail. FINISHING HAMMERS have smooth faces and range in weight from 7 to 16 ounces. Curved and rip claws are available. A general purpose hammer that could be used for both framing and finish work would have a smooth face and a 16-ounce head.

SCREWDRIVERS are necessary for both Phillips and slotted screws. Various sizes of each screwdriver are also necessary to fit all the screws used in construction. Standard screwdrivers are sized by the length of the blades and the size of the tips. Phillips screwdriver sizes refer to the size of the tips and are numbered from 0 to 4. A Number 2 Phillips screwdriver is the most common and will fit most Phillips screws, but other sizes will eventually be necessary to insure a proper fit on all screws.

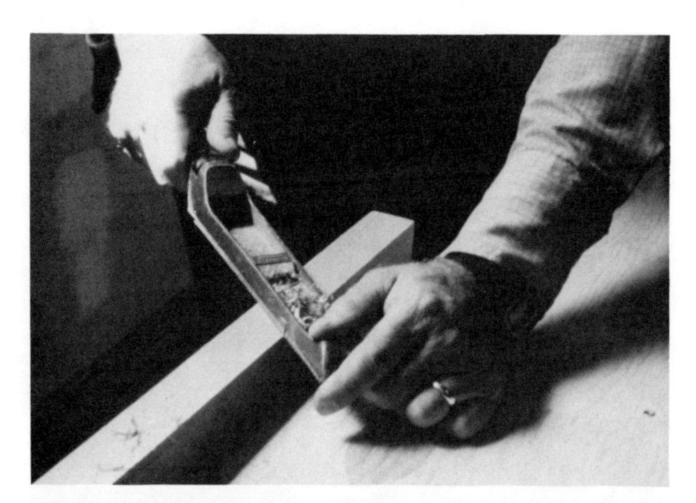

Photo 17. This tool works much like the plane.

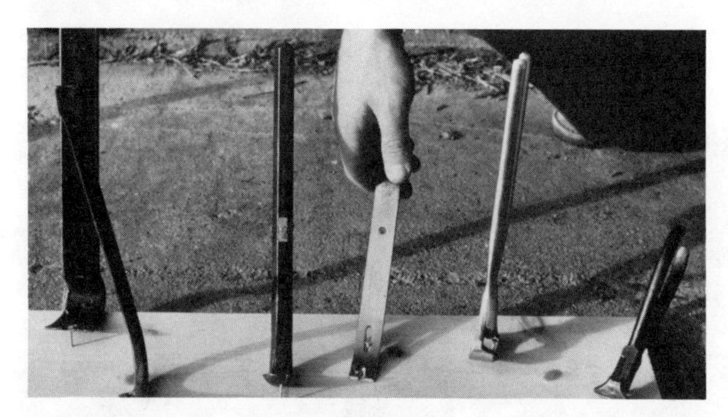

Photo 18. Above are various types of nail-pulling tools. Needless to say, it is highly recommended that all novice owner builders have a complete assortment as shown here. They will be well used. From left to right: large flat bar (wonder bar), small crow bar, cat's paw, small flat bar and a large and small set of nail clippers.

Photo 19. Shown here are some large wrecking and timber moving tools. It is a good idea to have the full assortment at the job site. From left to right: a hand winch (com-a-long), a sledge and a large crowbar.

Pliers and Wrenches

PLIERS are needed to hold steel pieces and fasteners during assembly.

For careful construction, nuts and bolts should have wrenches of appropriate size; however, a high quality ADJUSTABLE WRENCH is very valuable in holding a variety of pieces.

Smoothing Tools

The BLOCK PLANE is necessary for jobs such as shaving small amounts from a door, window, or a finish piece of wood. While power planes are replacing the larger sizes of hand planes, the control necessary for small adjustments and shaping makes it important to always have a block plane handy. (Photo 17.)

Several files of various shapes, sizes and coarseness are desirable. The most universal file for wood is a 4-IN-1 FILE. This file has 4 different surfaces which are quickly available in one tool, and it is easily carried in a tool belt. One side has coarse and fine flat surfaces. The other side has coarse and fine curved surfaces.

Prying and Wrecking Tools

The RIPPING BAR, also called a wrecking bar or pry bar, is used to pull previously nailed pieces apart, withdraw nails and lever pieces into place.

A FLAT BAR is used to remove trim and molding, separate previously nailed framing pieces, and withdraw nails. It can be driven between pieces with a hammer and pried straight back or to the side to separate pieces.

The NAIL CLAW, also called a CAT'S PAW (Photo 18), is driven into the wood under the head of a nail to start to pry it out. Further prying is done with a hammer, flat bar or wrecking bar. Both novices and pros make mistakes, so have one of these in your tool box.

SLEDGE HAMMERS (Photo 19) are necessary to drive stakes, wreck or move framing.

Boring Tools

Most holes are drilled with ELECTRIC DRILLS today, and some drill motors are battery powered. Bit braces and hand drills are useful only when electricity is not available or convenient. The type of hole necessary determines the bit that is used. For holes up to ½ inch in diameter, machine or twist bits are frequently used. These bits are useful for metal or wood cutting. For larger holes, spade or power bore bits can be used. However, to cut a truly round hole, remove chips quickly, and drive easily, a power auger bit is used.

POWER AUGERS are used with large drill motors at lower speeds. When very accurate holes with crisp edges are needed, as for exposed dowels or plugs, a craftsperson can use FORSTNER BITS. When screw heads must be flush or below the surface, holes are first drilled with a countersink (Photo 20).

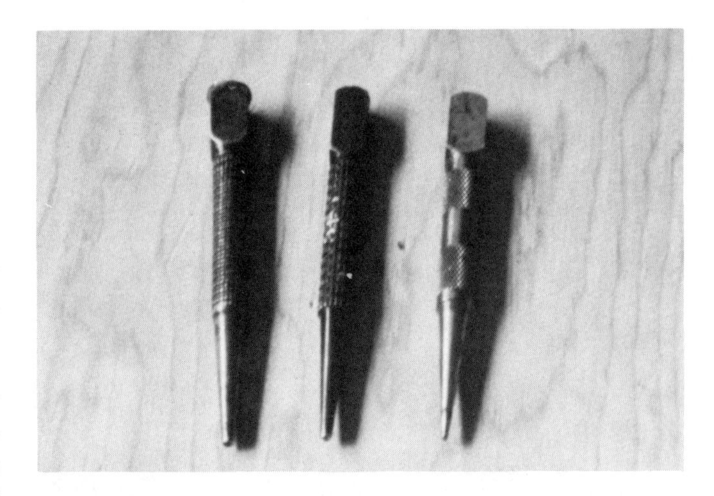

Photo 20. Nail sets or nail punches are available in different sized tips to countersink different sized nails.

Photo 21. Metal legs for saw horses are easy to use and more secure than wooden legs.

Power Tools

The CIRCULAR SAW (Photo 22–Photo 27) is usually one of the first power tools purchased. It is necessary to efficiently cut framing pieces. Since the saw will be used frequently it is important to choose a high quality tool that feels comfortable. The blade must always be sharp to prevent overloading the tool, damaging the work or causing a kickback toward the operator. To keep a sharp blade on the tool, some craftspeople choose carbide tipped blades which keep their keen edge longer. However with the rough service sometimes demanded of a blade, you may choose to have many less expensive steel blades to change frequently and have some at the saw shop for resharpening. Resharpening costs for steel blades are less.

Most Common Mistake

Setting a board *between* two supports to be cut. As the board droops during the cut it binds the blade and causes the saw to kick back.

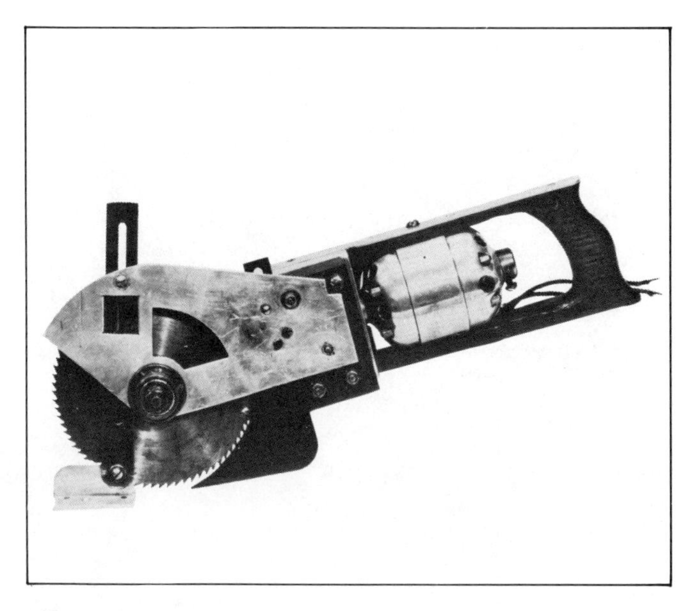

Photo 22. Circular saws have come a long way from this original one.

Photo 23. The power circular saw is perhaps the most commonly used power tool during the housebuilding process. Be sure to choose one that will last and that feels comfortable to you. The model above is referred to as a "sidewinder" as opposed to a "worm drive" model. Try them both and see which feels more comfortable.

Photo 24. With the proper masonry bit, power circular saws can be used to cut concrete block and brick.

Photo 25. A worm drive saw with blades. Note how dull blades are separated from the sharp blades on plywood boards. Have a good assortment of blades, both carbide tipped and standard.

Photo 26. Most circular saws can be adjusted so that the saw is at an angle to the cutting platform. This allows you to bevel cut the board up to 45°.

Photo 27. This 5½-inch circular saw is good for cutting paneling, siding and trim. Not necessary, but useful if you can afford it.

The SABRE SAW (Photo 28) is a portable jigsaw used for cutting a variety of materials and shapes. Blades are available for cutting wood, plywood, metal and plastics. The tool is useful for making cutouts in the center of building materials, including sheets of plywood and laminated countertops. Because the sabre saw has a very small blade, it is also very useful in making curved cuts.

Helpful Hints and Most Common Mistakes

☐ Finished surfaces such as plastic laminates can be scratched by the base of the sabre saw. Covering the base of the saw with tape will prevent damage.

☐ Be careful not to pinch a finger between the blade chuck and the body of the tool while it is operating.

☐ Choose the appropriate saw blade to insure a fast and smooth cut.

Photo 28. A jigsaw is very useful at most building sites. It can cut curves and angles and fancy scrolls.

ELECTRIC DRILLS (Photo 29–Photo 32). The most universal drill for a carpenter is a ⅜-inch, variable speed, reversing drill. This tool allows bits up to ⅜ inch to be used for drilling. Occasional larger holes can be done with wide power bits with smaller shanks. For continued large hole boring and heavy-duty use, a more powerful ½-inch drill is more appropriate. As chuck capacities go up, drills are designed to turn more slowly to produce more torque with less likelihood of operator injury if the bit should jam.

It is also possible to use this ⅜-inch drill with a screwdriver tip (usually Phillips) in the chuck to make it a power screwdriver. The variable speed trigger plus forward and reverse features provide ample control.

Photo 29. ¼-inch drills are handy for smaller projects, but usually ⅜-inch or ½-inch drills are used for larger jobs.

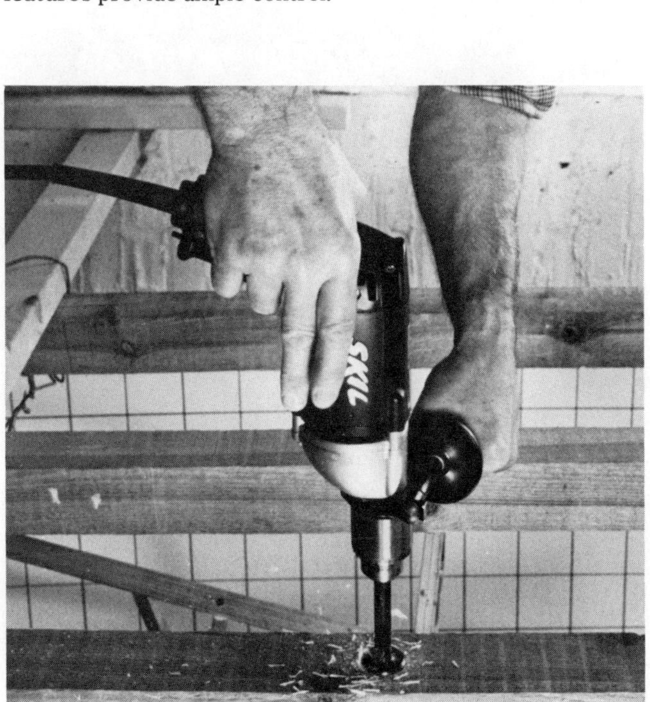

Photo 30. A ⅜-inch, heavy-duty drill is a must for anyone planning to do their own plumbing and electricity.

Photo 31. Cordless, rechargeable drills are handy during all stages of construction.

Photo 32. Small cordless screwdrivers are a new tool on the market. They come in very handy in many phases of building.

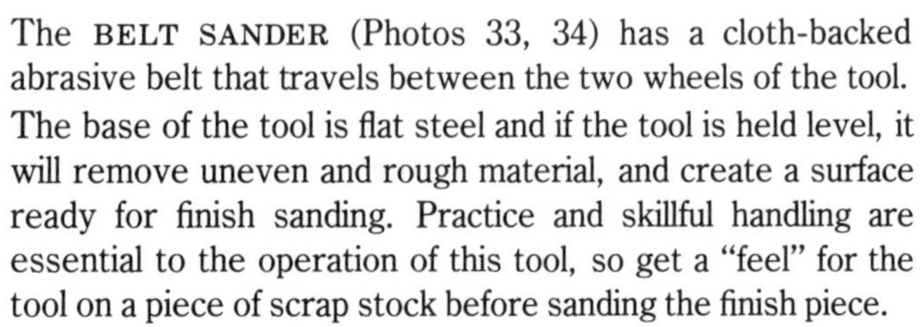

The BELT SANDER (Photos 33, 34) has a cloth-backed abrasive belt that travels between the two wheels of the tool. The base of the tool is flat steel and if the tool is held level, it will remove uneven and rough material, and create a surface ready for finish sanding. Practice and skillful handling are essential to the operation of this tool, so get a "feel" for the tool on a piece of scrap stock before sanding the finish piece.

Sanding should always be done by using progressively finer grits of abrasive belts and paper. Each finer grit removes the scratches of the previous grit until the desired finish is achieved.

Helpful Hints and Most Common Mistakes

☐ Belt sanders must be held level to achieve good results. Often it is necessary, when doing surfaces with pieces cut out of them, to imagine a completed surface and hold the belt sander to the imagined plane. Dipping the sander into the openings or applying more pressure to the corners will cause gouging of the work.

☐ Belt sanders must be kept moving from area to area or the belt will dig in.

☐ Sanding belts are usually directional and putting belts on backwards tears the seams. Some non-directional belts, however, allow additional sharp usage by reversing them.

Photo 33. A heavy-duty belt sander like this is usually not needed in a typical housebuilding process. It is used for large sanding jobs where the weight of the machine is needed to make sanding easier.

Photo 34. The belt sander is used for larger sanding jobs. Wearing safety glasses as shown is recommended whenever you are using any power tool. If you wear glasses, use hardened glass or plastic.

The FINISHING SANDER or PAD SANDER (Photo 35) is useful in smoothing away tool marks and wood fillers. Some are available with orbital and "in-line" motions. Orbital action removes stock more quickly, while "in-line" action allows sanding with the grain of the wood for a smoother finish. Finishing sanders have a cushioned base of felt or rubber, and are designed for smoothing. To remove a lot of material and make a flat surface the belt sander is used.

Helpful Hints and Most Common Mistakes

☐ Different level wood joints sanded to flat with a finish sander will be smooth, but not flat and straight.

☐ Trim wood should be sanded with a grit of sandpaper as high as 220 to remove machine marks that will be evident when the stain is applied; but sanding with too fine a grit can burnish the wood cells closed and cause a blotchy stain effect. The highest sanding grits are used between coats of finish.

Photo 35. Finishing sanders are used for lightweight sanding such as furniture refinishing. Many people use them when sanding dry wall as well.

Additional Tools

ROUTERS (Photo 36) are basically edge-forming tools, but with additional guides they can be used to cut grooves, rabbets, dadoes and mortises for hinges. A collection of bits for routers can frequently exceed the cost of the router itself. Bits are available from high-speed steel or carbide-tipped steel. When pilot tips are part of the cutter, steel bits have a fixed pilot and carbide cutters usually have ball bearing pilots. The carbide cutters stay sharp longer, cut more cleanly, and the ball bearing pilots do not burn the wood. For infrequently used profiles, the steel bits allow a greater variety of cutters at a more economical price.

Helpful Hints and Most Common Mistakes

☐ The base of the router must be kept flat to the board during use to insure profile accuracy.

☐ Since the cutter revolves in a clockwise direction (as seen from the top) the router should be moved from right to left to move the cutter into the work. A full-depth cut from left to right will cause the cutter to "climb" the wood and start to self-feed; however, a shallow precut from left to right followed by a full depth cut from right to left will prevent splintering on difficult woods.

☐ Rapid movement can cause splintering of the wood. Slow movement with solid steel pilots will cause the cutter to burn the wood. Practice on a scrap of wood to develop the proper technique.

Photo 36. A router is an excellent finishing tool. It can round edges, make different types of grooves, finish formica, etc.

Photo 37. A heavy-duty extension cord with a four-way junction box is essential.

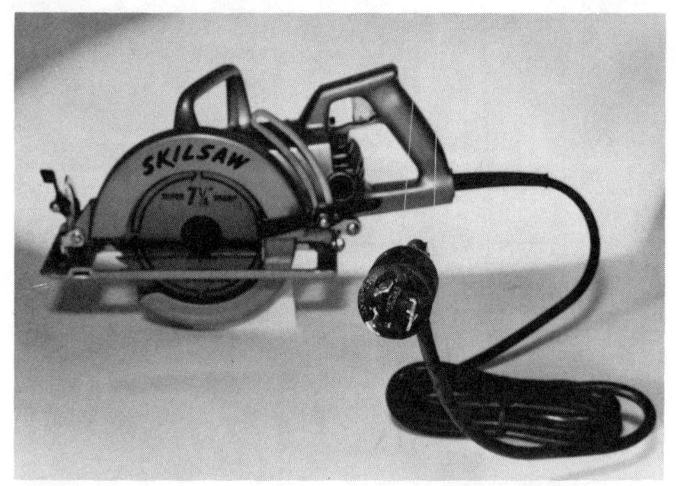

Photo 38. Many professionals use a locking plug as shown here. This male plug takes a corresponding female adapter. Once inserted and twisted, the plugs are locked together and will not pull apart. You may want to consider changing your regular plugs on your commonly used tools to this style.

HEAVY-DUTY EXTENSION CORD (Photos 37, 38). Along with the purchase of your first portable power tool should come the purchase of a heavy-duty (large wire diameter and heavily insulated) extension cord. When power must run a long distance to the tool, it loses effectiveness with small cords and can damage the tool. Nor will a small cord stand up to the wear and tear of use on a job site. Local safety requirements may dictate that the cord and tool be outfitted with twist-locking plugs, as shown, for secure connections.

A heavy-duty ½-INCH DRILL is essential for those who do their own plumbing and electrical work, and very desirable for boring long or frequent large holes for bolts and other fasteners. A right angle drill allows the operator to get in between framing members easily.

Helpful Hints and Most Common Mistakes

☐ Many bits used for drilling large holes have self-feeding pilot screws that assist in drawing the cutting edges into the wood. Care should be taken to keep face and hands clear of projecting handles. Frequently, injuries occur when a bit binds in a hole and the tool is wrenched from the operator's hands.

☐ Bit extensions and long bits are available to drill through several studs in one pass.

☐ When drilling deep holes, withdraw the drill bit occasionally to clear the chips.

The RECIPROCATING SAW (Photo 39) is similar to a sabre saw but much heavier in construction. It is the surgery tool of house construction. It is used to carve out holes for vents, cut openings for lights and other objects, notch studs for plumbing or electrical wires, and modify existing framing. Metal blades will cut nails, pipe, conduit and other pieces. It is an all-purpose saw, and its long-bladed snout allows the operator to reach into otherwise inaccessible areas.

Helpful Hints and Most Common Mistake

☐ Reciprocating saws bounce less and break fewer blades if the saw guide of the tool can be held tightly to the material.

POWER MITER SAWS (Photo 40) have largely replaced hand miter boxes in the woodworking shops and at construction sites. Nicknamed "chopsaws," they easily cut through material as thick as framing stock, bringing cabinet shop accuracy to the field, for clean miters on casings, handrails and all moldings. By setting a block to butt against, power miter boxes can be used to cut many pieces to the same length or angle without repeating measurements.

Helpful Hints and Most Common Mistake

☐ The power miter saw should not be overused. Straight cuts for framing that do not require this accuracy should be made with the circular saw at the spot they will be used. This saves time and handling effort.

A POWER PLANE quickly smoothes surfaces and trims materials to size. It is operated in the same manner as a hand plane, but since the rotating cutters smooth the surface, the tool moves more easily and the motion is not affected by knots or irregular grain.

Helpful Hints and Most Common Mistakes

☐ The blades are exposed on the bottom of a planer and subject to damage if they come in contact with a hard surface, while running or not.

☐ When starting cuts, pressure should be applied more to the front of the tool. When completing cuts, pressure should be applied more to the rear of the tool.

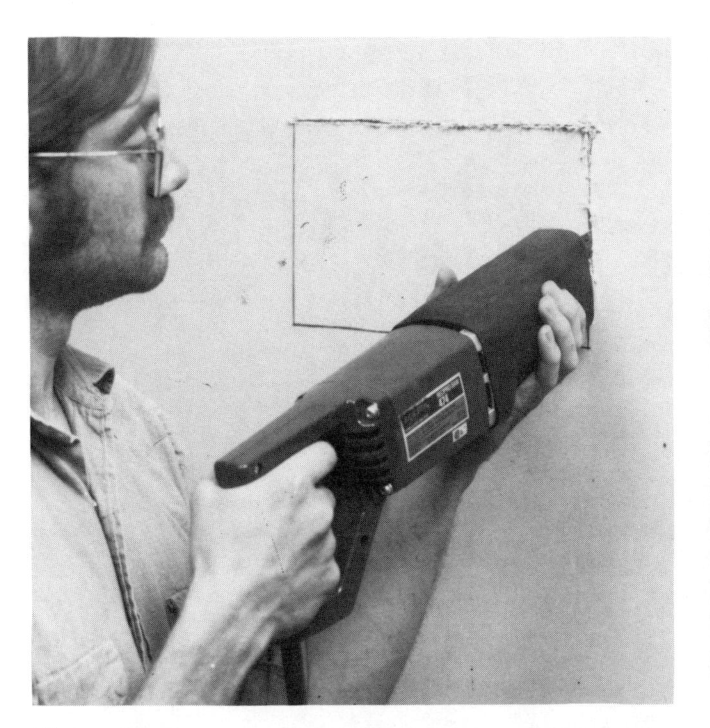

Photo 39. The reciprocating saw has a knife-like blade that protrudes from the front of the saw and cuts in a back-and-forth direction. When you need this tool, nothing else will do.

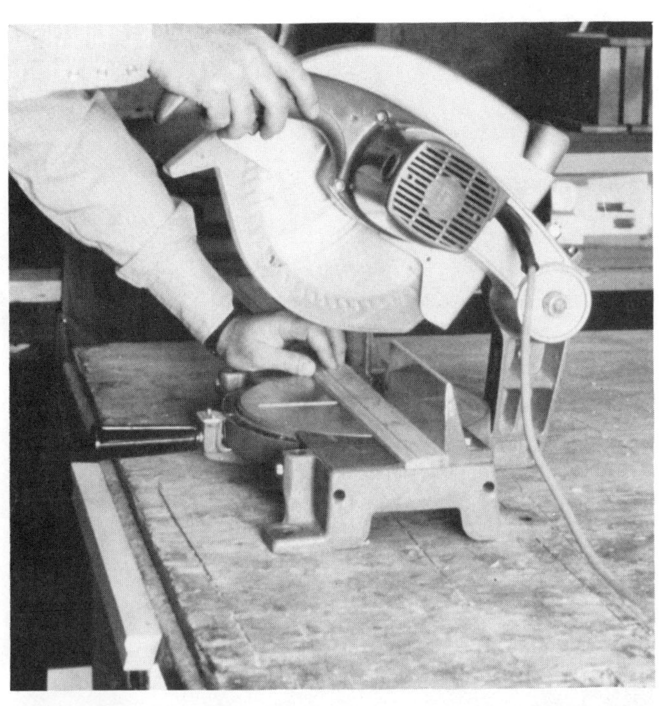

Photo 40. A power mitre saw (chop saw) is a professional tool used for finishing work. You may want to rent one or buy one if you are building a large custom house.

Photo 41. Pneumatic tools, powered by compressed air, make many jobs quicker and easier. A professional tool, they can usually be rented at certain phases of construction.

Photo 42. A radial arm or contractor's saw is useful both during the framing and finishing stages. It is expensive and not necessary for a one-house project, but very useful if you can afford it.

PNEUMATIC TOOLS (Photo 41). Pneumatic (air driven) tools such as staplers and nailers are available to increase production. When a good hand rhythm is achieved, work goes quickly and smoothly. To operate pneumatic tools, the tool must be connected with an air hose to a compressor. This air hose "umbilical" is restrictive and of questionable value, especially considering the investment necessary for a single house.

Some contractors find pneumatic tools useful, however, for multiple units and large square footage jobs such as siding and roofing, and some framing sub-assemblies. The decision to use pneumatic tools or not depends on the job conditions, craftspeople, project design and total square footage. Pneumatic tools are enjoyable to use, less tiring, fast and effective, but not essential.

RADIAL ARM SAW (Photo 42). One of the first stationary power tools for many people is the radial arm saw. It is useful on the construction site for cutting beams, repetitious framing pieces, angles and compound angles. This saw is basically designed to draw the blade across (crosscutting) the wood. Ripping (cutting with the grain) in the long direction of a board is usually done with a portable circular saw or table saw.

Most Common Mistake

☐ Since the radial arm saw is a stationary tool the work must be brought to the tool. Making too many cuts with this saw, especially when work must be transported from another floor, is very time-consuming. Unless there is unusual thickness or angles, or a need for extreme accuracy is involved, the circular saw should be used for framing.

The TABLE SAW (Photo 43–Photo 46) excels at cutting sheet stock such as plywood, and cutting the long way of a board (ripping). It is more difficult to cut across the grain (crosscut) on a table saw since the whole board must be moved across the blade. For carpenters who will be doing their own cabinets and built-ins, a table saw will be very useful.

Most Common Mistake

☐ The table saw is outfitted with two guides to use in cutting stock, a miter gauge used at angles to the blade and a rip fence which is parallel to the blade. When cutting stock, choose the guide that will allow the longest edge of the stock to be against the guide. Do not have the stock touching both the miter gauge and the rip fence while cutting, or binding and kickback may occur.

Photo 43. Table saws are excellent for doing trim and finishing and cabinet work. They are not a required tool for the one-time owner builder, but are useful if you can afford it. The one above is a smaller homeowner's model. It should be all that is needed for a one-house project.

Photo 44. Large bench saws are not essential, but very useful if available.

Photo 45. Table saws can use dado blades to make grooves like these. Many other blades are also available.

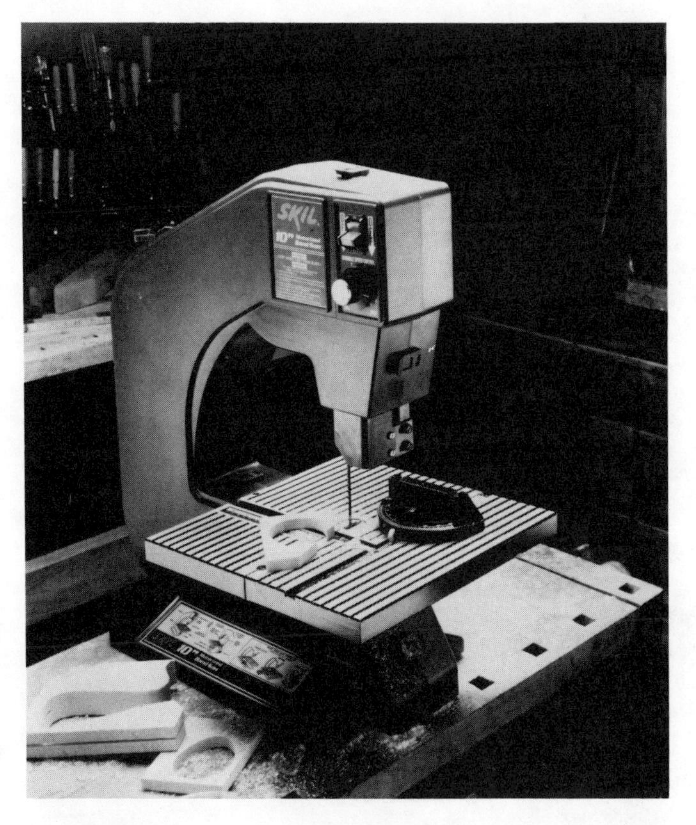

Photo 46. Band saws are used mostly in finishing and sculpturing work. They are useful for cutting angles and curving pieces of wood. You can probably get by with the hand-held equivalent, the jig or sabre saw.

Specialized Tools

Specialized tools are not part of everyone's toolchest. They have specific purposes and are seldom needed, but essential when they are needed. You may want to buy, borrow or rent these tools. Also a tour of the local tool rental shop may find a tool you would like to have occasionally. A FOUNDATION STAKE PULLER saves time and money, for instance, but is a one-day need.

Other tools to consider are the com-a-long and hardwood floor nailer. A COM-A-LONG is a device with a cable and ratcheting handle used to pull framing sections into square and plumb. The FLOOR NAILER holds and aligns the nails along the edge of hardwood flooring. It drives a nail when struck with a hammer or mallet.

Additional hand tools with specialized purposes are the SHINGLE RIPPER for removing nails during roofing repairs, RABBET PLANE for shaping rabbet cuts in wood and a NAIL SHOOTER for driving nails in difficult spots.

Hundreds of tools also exist that are used for special trades only. Drywallers, carpet layers, tilesetters, roofers, plumbers, electricians and others have occasional need for more specialized tools. To become familiar with all the tools available, visit the trade supply distributors. While a one-time need will not justify a complete extension of your tool collection for every trade, some special tools may be helpful. (Photos 3-47, 3-48).

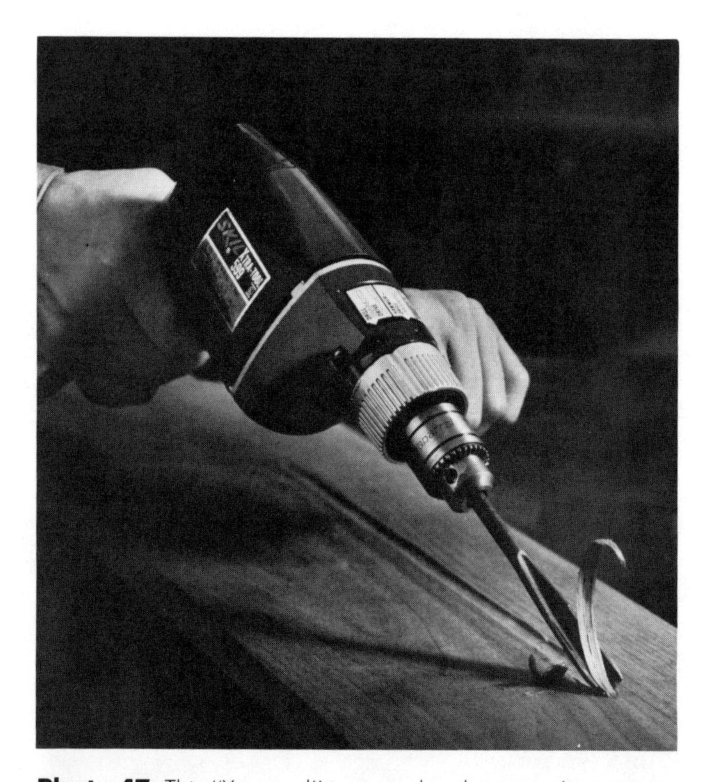

Photo 47. This "Xtra-tool" is several tools in one. It can serve as an electric screwdriver or drill, as well as a hammer-drill, or hammerchisel. It can drill into brick or concrete, chisel or gouge wood, and be used to remove paint or linoleum as well as chip concrete. A useful tool.

Photo 48. Shop vacuum cleaners are useful, especially during construction stages.

Tool Checklist

	Tool		
☐	Combination stone	☐	$ ______
☐	Metal files	☐	$ ______
☐	20- or 25-foot tape	☐	$ ______
☐	Plumb bob	☐	$ ______
☐	Chalkline	☐	$ ______
☐	Pocket knife	☐	$ ______
☐	Utility knife	☐	$ ______
☐	Carpenter's pencil	☐	$ ______
☐	Carpenter's square	☐	$ ______
☐	Combination square	☐	$ ______
☐	2-foot level	☐	$ ______
☐	Crosscut saw	☐	$ ______
☐	Rip saw	☐	$ ______
☐	Coping saw	☐	$ ______
☐	Hacksaw	☐	$ ______
☐	Wood chisels	☐	$ ______
☐	Cold chisels	☐	$ ______
☐	Diagonal cutting pliers	☐	$ ______
☐	Nail sets	☐	$ ______
☐	Hammer	☐	$ ______
☐	Screwdrivers	☐	$ ______
☐	Phillips screwdrivers	☐	$ ______
☐	Pliers	☐	$ ______
☐	Adjustable wrench	☐	$ ______
☐	Block plane	☐	$ ______
☐	4-in-1 file	☐	$ ______
☐	Flat bar	☐	$ ______
☐	Nail claw	☐	$ ______
☐	Twist bits	☐	$ ______
☐	Spade bits	☐	$ ______
☐	Circular saw	☐	$ ______
☐	Sabre saw	☐	$ ______
☐	Belt sander	☐	$ ______
☐	Finishing sander	☐	$ ______
☐	Heavy duty extension cord	☐	$ ______
☐	Bench grinder	☐	$ ______
☐	50- or 100-foot tape	☐	$ ______
☐	Builder's level or transit	☐	$ ______
☐	Sledge hammer	☐	$ ______
☐	Bit brace	☐	$ ______
☐	Hand drill	☐	$ ______
☐	Power auger bits	☐	$ ______
☐	Forstner bits	☐	$ ______
☐	6-foot level	☐	$ ______
☐	Framing hammer	☐	$ ______
☐	Assorted files	☐	$ ______
☐	Ripping bar	☐	$ ______
☐	Countersink	☐	$ ______
☐	Router and bits	☐	$ ______
☐	½-inch right angle drill	☐	$ ______
☐	Reciprocating saw	☐	$ ______
☐	Power miter saw	☐	$ ______
☐	Power plane	☐	$ ______
☐	Com-a-long	☐	$ ______
☐	Rabbet plane	☐	$ ______
☐	Nail shooter	☐	$ ______
☐	Aviation snips	☐	$ ______
☐	Laser level	☐	$ ______
☐	Foundation stake puller	☐	$ ______
☐	Hardwood flooring nailer	☐	$ ______
☐	Shingle ripper	☐	$ ______
☐	Battery powered drill	☐	$ ______
☐	Pneumatic nailer	☐	$ ______
☐	Pneumatic stapler	☐	$ ______
☐	Radial arm saw	☐	$ ______
☐	Table saw	☐	$ ______

Glossary

ACTIVE SOLAR. Use of solar hardware, panels, plumbing, gauges, etc., to use the sun's energy to heat hot water, or the home.

BBC. Basic Building Code.

BERM. Constructing the house so that it is partially or totally underground, or covered with earth.

CLIVUS MULTRUM. A brand of composting toilet.

COOLTH TUBE. A tube run several feet underground, through which air is carried into the home. The air is cooled by its contact with the pipe, which is in turn cooled by its contact with the soil.

DEED. A document executed under seal and delivered, to effect a conveyance of real estate.

EASEMENT. A right held by one property owner to make certain kinds of use of the property of another; i.e., to cut timber, mine minerals, or travel across.

ESCROW. A contract or money deposited with a third party, to be held until the fulfillment of certain conditions; i.e., the closing of the sale of property.

GPM. Gallons per minute of water flow.

GRAY WATER. Waste water from sinks, showers, tubs and washing machines, not including human waste.

HEAT TAPE. Tape carrying an electric current. Wrapped around water pipes, it heats and protects them from freezing.

INSOLATION. Solar radiation received over a given area.

JOISTS. The main floor framing members of a house.

LEACH FIELD. Area underground through which run perforated drain pipes from the septic tank, carrying effluent wastes.

LIEN. A legal claim of one person on the property of another person, for the payment of a debt or the satisfaction of an obligation.

LOAD BEARING CAPACITY (SOIL). The ability of soil to withstand the loads applied to it by a building before sinking or shifting.

MECHANIC'S LIEN. A lien imposed by anyone who has improved or worked on a house or property, or supplied materials for either.

MORATORIUM. A temporary cessation in the issuance of building permits.

MORTGAGE. The transfer of real property to a lender as a guarantee for a loan.

PASSIVE SOLAR. Use of site orientation, building design, south-facing windows, and thermal mass, to enable a house to collect and store the sun's energy for heating.

PELTON WHEEL. A small water wheel that can be used to generate electricity from small creeks or streams.

PERCOLATION TEST. A test of the soil to ascertain if it is porous enough to adequately drain and filter a septic tank leach field.

PERFORMANCE BOND. Money that guarantees certain work will be done in a prescribed and previously agreed upon fashion.

PERMIT. Legal permission to build, as granted by a local government regulatory agency, usually the building permit department.

pH. A measure of the acidity or alkalinity of water, given as 7 for neutral, increasing with increasing alkalinity and decreasing with increasing acidity.

PRELIMINARY TITLE REPORT. A report, usually prepared by a title company, describing the clearness of the title and whether there are any liens, encumbrances, outstanding notes, easements, or other complications that could make ownership or construction problematic.

SETBACK. A required distance a building must be from various property lines.

SOIL PROFILE. A description of the soil, its composition, and load bearing capacity.

SUBORDINATION. A legal agreement whereby one lender agrees to collect his loan after another lender has collected his, should the borrower default on the loan.

SUBSOIL. The layer of earth, often sterile, below the top layer of soil.

TITLE. Legal right to the possession of property.

TOPSOIL. The top few inches of soil that is made up of decaying organic matter, conducive to growing plants.

UBC. Uniform Building Code.

VARIANCE. An official permit to do something normally forbidden by building or zoning regulations.

VOUCHER SYSTEM. A payment system used by lending institutions for the disbursement of construction funds. Suppliers, contractors and workmen are paid by certificates and vouchers which are exchanged for money by the lending institution once the work has been satisfactorily completed and inspected.

WATER TABLE. The level below the surface at which underground water is located.

Appendices

ALL COST ESTIMATE charts should be used with reservations. Inflation and the details of each project cause great variances between chart projections and real costs. Also, charts are geared to non-custom homes; custom houses are always more expensive and harder to estimate. The following charts may be of some use to you both in your estimate projections and as examples of how estimates are done.

I suggest that you discuss your estimate with a professional builder or contact one of the owner-builder schools and have one of their consultants check it out before you begin. See *Consulting Service*, on page 145.

Appendix A

This chart was prepared by a major west coast bank. The house it uses for a model is described below. The following chart then tells how much each process cost, changes in cost from past years, percentage of total cost, and cost per square foot.

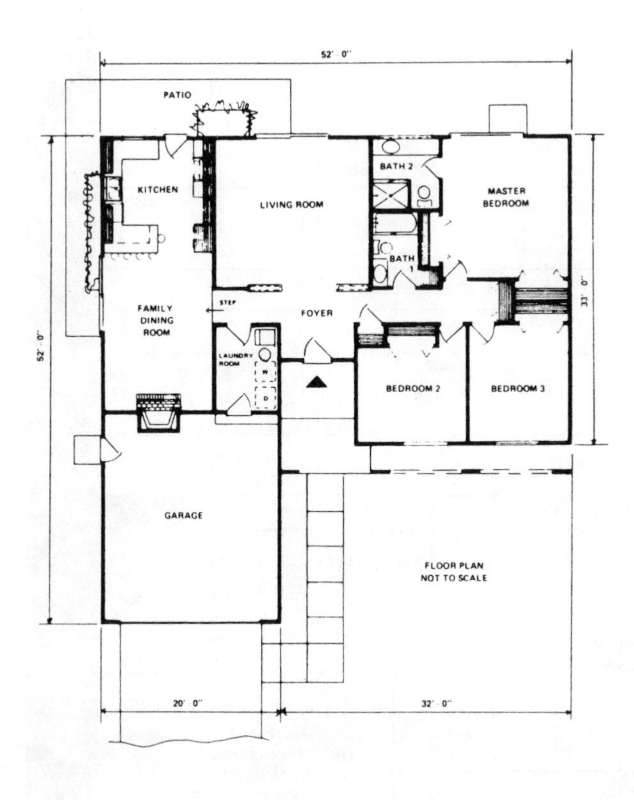

COST STUDY

PERSPECTIVE
FLOOR PLAN
GENERAL DESCRIPTION

Standard quality "semi-custom" (non tract) residence (1,570 s.f.) consisting of 3 bedrooms, 1-¾ baths, living room, kitchen, family-dining room, laundry room, attached 2 car garage (446 s.f.), and patios, driveway and walks (837 s.f.).

FOUNDATION:	Perimeter reinforced concrete and interior piers
FLOORS:	2:4:1 (1-1/8") plywood over 4 x 6 girders @ 4'-0" O.C. Mesh reinforced 4" concrete slab over rock base and membrane at kitchen, family-dining room
WALLS:	Stucco with integral color coat over 2 x 4 studs @ 16" O.C.
PARTITIONS:	Painted gypsum wallboard
INSULATION:	Ceiling (R-19), wall (R-11), floor (R-11)
WINDOWS AND SLIDING DOORS:	Anodized aluminum frames with one-half screens, single glazing
DOORS:	Pre-hung masonite and wood
ROOFING:	Medium taper split cedar shakes, 4:12 Hip
CEILINGS:	Simulated sprayed acoustical plaster and painted gypsum wallboard
FINISH FLOORING:	Sheet vinyl in kitchen, family-dining room, bathrooms, laundry room, foyer; carpeting elsewhere
FIREPLACE:	Prefabricated metal with brick facing and hearth
CABINETS:	Mill built with laminated plastic and cultured marble tops
HEATING:	Gas fired 100,000 BTU forced air with high wall registers
PLUMBING:	Copper water lines, ABS waste lines
ELECTRICAL:	Romex, 125 amp underground service
BUILT-INS:	Range with hood and double ovens, garbage disposal and dishwasher

COST STUDY

JUL 1981

STANDARD QUALITY SINGLE FAMILY RESIDENCE
SAN FRANCISCO AREA

Division		Item	$ COST AS OF 7/1/81	% CHANGE SINCE 4/1/81	% CHANGE SINCE 7/1/80	% OF TOTAL DOLLARS	$ COST PER S.F.
PRELIMINARY AND GENERAL CONDITIONS	1.0	.1 Permits, plan checking, temporary power, water, portable toilet, debris box	1,155[b]	+ 7.9	+ 17.4	1.7	.74
		.2 Final clean-up (allowance)	325	0	0	.5	.21
SITE WORK	2.0	.1 Site preparation and excavation	850	0	+ 13.5	1.3	.54
		.2 Flatwork (driveway, patio, walks)	1,685		+ 11.7	2.5	1.07
CONCRETE	3.0	.1 Foundations, slabs, piers	2,672	+ 2.6	+ 14.3	4.0	1.70
MASONRY	4.0	.1 Brick hearth and face veneer at fireplace	477	+ 6.5	+ 11.2	.7	.30
METAL	5.0	.1 Rough hardware	277	0	a	.4	.18
		.2 Finish hardware (allowance)	150	0	0	.2	.10
WOOD AND CABINETRY	6.0	.1 Rough lumber	5,853	− 1.3	− 2.0	8.8	3.73
		.2 Finish lumber	339	+ 4.6	+ 14.9	.5	.22
		.3 Rough carpenter labor	5,575	+ 7.9	+ 11.2	8.3	3.55
		.4 Finish carpenter labor	1,115	+ 7.9	+ 11.2	1.7	.71
		.5 Countertops (cultured marble and laminated plastic)	976	0	+ 9.9	1.4	.62
		.6 Cabinets	2,371	a	+ 9.6	3.5	1.51
THERMAL AND MOISTURE PROTECTION	7.0	.1 Insulation, weather stripping, thresholds	1,482	0	+ 8.7	2.2	.94
		.2 Roofing (medium shakes)	3,600	a	− 2.9	5.3	2.29
DOORS AND WINDOWS	8.0	.1 Doors	1,271	+ 1.5	+ 3.0	1.9	.81
		.2 Garage door	281	a	a	.4	.18
		.3 Aluminum windows, glass sliding doors, all with screens	969	+ 2.8	+ 11.3	1.4	.62
FINISHES	9.0	.1 Stucco	4,199	+ 10.7	+ 18.0	6.2	2.67
		.2 Gypsum wallboard, ceiling acoustical spray	2,859	+ 2.5	+ 2.5	4.2	1.82
		.3 Resilient flooring (allowance)	1,300	0	0	1.9	.83
		.4 Carpeting (allowance)	1,550	0	+ 10.7	2.3	.99
		.5 Painting	2,595	+ 5.6	+ 9.5	3.8	1.65
SPECIALTIES	10.0	.1 Shower and tub enclosures	287	0	+ 9.1	.4	.18
		.2 Prefabricated fireplace	611	a	+ 5.3	.9	.39
		.3 Bath accessories (allowance)	400	0	0.0	.6	.25
APPLIANCES	11.0	.1 Built-ins (allowance)	1,100	+ 10.0	+ 10.0	1.6	.70
	12.0/13.0/14.0	(No items under these divisions)					
MECHANICAL	15.0	.1 Heating and sheetmetal	2,579	+ 7.8	+ 10.8	3.8	1.64
		.2 Plumbing, including sewer connection	5,415	+ 5.4	+ 10.2	8.1	3.45
ELECTRICAL	16.0	.1 Wiring	2,164	+ 11.5	+ 12.6	3.2	1.38
		.2 Fixtures (allowance)	700	0	+ 7.7	1.0	.45
		SUB-TOTAL	57,182	+ 3.8	+ 7.8	84.7	
	A.	Insurance: Workers' Compensation, Social Security, Unemployment	1,405	+ 7.9	+ 11.2	2.1	.89
	B.	Overhead and profit 15%	8,577	+ 3.8	+ 7.8	12.8	5.47
	C.	Plans and Specifications	301	0	0	.4	.19
		TOTAL CONSTRUCTION COST	67,465	+ 3.9	+ 7.9	100.0	42.97

SUMMARY

	AREA	S.F. COST	TOTAL
House	1570 S.F.	37.49	$ 58,859
Garage	446 S.F.	15.52	6,921
Patios, Driveway, Walks	837 S.F.	2.01	1,685
TOTAL CONSTRUCTION COST			$67,465

NOTES:
(a) Denotes change of less than 1%.
(b) Miscellaneous municipal fees & taxes not included.

These work sheets can be used in your own project. The numbers used here are for a very small home (959 sq. ft.) and is used only as an example. It will, however, give you a sense of how estimates are arrived at. I suggest adding a 20–25% "fudge factor" to all estimates.

COST SUMMARY
(Small House)

A.	Foundation	$ 450.65		
B.	Plywood	$ 1,639.60		
C.	Doors	$ 585.00		
D.	Insulation	$ 462.90		
E.	Framing Lumber }	$ 2,458.06		
F.	Trim }	$ —		
G.	Rough Hardware	$ 231.60		
H.	Finish Hardware	$ 209.00		
I.	Miscellaneous	$ 400.00		
		SUB-TOTAL$_1$	$ 6,436.81	Permit $500
		TAX	$ 418.39	Site Prep 250
				Plans 650
		TOTAL$_1$	$ 8,795.20	Insurance 540
				1,940
J.	Carpentry Labor	$ 6,534.11		
		TOTAL$_2$	$15,329.31	
K.	Sub-Contractor & Supplier	$18,096.00		
L.	General Contractor's profit & Overhead	$ 6,685.00	$33,425.31	
		GRAND TOTAL	$40,110.00	= $ 41.8 /S.F.

OPTIONS

1. Non Sub-contract costs	$ 5,942.86		
ADD TOTAL$_2$ ABOVE	15,329.31		
	GRAND TOTAL$_1$	$21,272	= $ 22.2 /S.F.
2. Non Sub-contract costs	$ 5,947.86		
ADD TOTAL$_1$ ABOVE	$ 8,795.20		
	GRAND TOTAL$_2$	$14,738.00	= $ 15.4 /S.F.

WORK SHEET

A. FOUNDATION:

1) Concrete -

a) Footing: 128 x ½ = 64

b) Wall: 128 x (3/2 x 1/2) = 96   = 160   161.50 / 27 = 6.0

c) Piers: 3 x 1/2 = 1.50

6½ Cu. Yards x $50/cu.yd. = $ 325.00

2) Rebar (#4 = 1/2") -

128 Lin. Ft. Wall x 3 + 10% Lap - 440 Ft.

440 Ft. x $0.18/Ft. = $ 79.20

3) Foundation Hardware and Piers -

a) Anchor bolts (½"x10") : 128 Ft. x 1/6 Ft.

20 + 8 = 28 Ft. x $0.60 = $ 16.80

b) Quik-ties (6"): 128 Ft. x 1/3 Ft.

43 + 4 = 47 x $.25 = $ 11.75

c) Wedges: 94 x $.10 = $ 9.40

d) Tie wire: 1 roll/500' = 1 x $2.50 = $ 2.50

e) Piers: 3 x $2.00 = $ 6.00

TOTAL $ 450.65

WORK SHEET

B. PLYWOOD:

1) Floor - 5/8" D.F. CDX T&G:

960 x $435/100 sq. ft. = $ 417.60

2) Roof - 1/2" D.F. CDX sq. edge:

1,280 x $320/100 sq. ft. = $ 409.60

3) Exterior Walls - 5/8" D.F. T 1-11:

4 @ (8') 128 x $525/1000 sq. ft. = $ 67.20

36 @ (9') 1,296 x $575/1000 sq. ft. = $ 745.20

TOTAL $ 1,639.60

C. DOORS:

1) 4 - 3' Flush H/C R.H. x 6'8" @ $55/ea. $ 220.00

2) 3 - 3' Flush H/C L.H. x 6'8" @ $55/ea. $ 165.00

(above pre-hung w/4½ jambs-casing 2 sides)

3) 2 - 3' Flush S/C R.H. x 6'8" @ $100/ea. $ 200.00

(above pre-hung w/4 5/8" jambs-casing inside)

TOTAL $ 585.00

D. INSULATION:

1) R-11: 870 sq. ft. x 19¢/sq. ft. = $ 165.30

2) R-19: 960 sq. ft. x 31¢/sq. ft. = $ 297.60

TOTAL = $ 462.90

E. FRAMING LUMBER:

1) FLOOR -

a) 2 x 4 sill: 4/20, 4/12

b) 4 x 4 post: 1/6

c) 4 x 10 beam: 2/20

d) 2 x 8 floor joist: 33/14, 36/12, 3/18, 7/20

2) WALL -

a) Studs (92¼) 222 + 22 = 244

b) 2 x 4 R.L.: 222 x 4 = 888 + 133 = 1,021 ft.

c) 4 x 12 headers: 1/20, /18, 1/14

3) CEILING & ROOF -

a) 2 x 6 clg. joists: 21/14, 21/12, 4/20

b) 2 x 8 rafters: 42/16, 2/20, 2/18

c) 2 x 8 backing: 3/12

d) 1 x 10 ridge: 2/20

F. TRIM:

1) Interior - 3/8" x 2¼" baseboard: 256 ft.

2) Exterior - 1 x 4 D.F. Select: 15/18, 4/14

G. ROUGH HARDWARE:

1) Nails: 400#/100 sq. ft. =	4	x $40/100#	= $	160.00
2) Ply-clips:	1	x $15/box	= $	15.00
3) Attic vents:	2	x $8.50	= $	17.00
4) Joist hangers:	6	x $0.40	= $	2.40
5) Foundation vents:	12	x $2.50	= $	30.00
TOTAL			$	231.60

H. FINISH HARDWARE:

1) Door Latch sets:	9	x $15.00 =	$	135.00
2) Cabinet pulls:	20	x $1.50 =	$	30.00
3) Towel Bars, etc.:	3	x $8.00 =	$	24.00
4) Medicine cabinet:	1	x $20.00 =	$	20.00
TOTAL			$	209.00

. MISCELLANEOUS:

Stake rental, staples, string, glue, const. adhesive, caulking, saw and blade, sharpening, #15 felt, temporary utilities, tool rentals.

TOTAL $ 400.00

J. CARPENTRY LABOR:

1) Foundation layout-
128 ft. Wall x 0.12 M.H./ft. wall = 15.36 man hours

2) Form & pour concrete-
128 ft. Wall x 0.67 M.H./ft. wall = 85.76 man hours

3) Floor & install sub-floor-
960 sq. ft. x 40 M.H./1,000 sq.ft. = 38.40

4) Frame walls -
222 ft.wall x 0.25 M.H./ft. wall = 55.50

5) Frame ceiling & rafters-
960 sq. ft. x 72. M.H./1,000 sq.ft. = 69.12

6) Sheath roof -
1,280 sq. ft. x 20 M.H./1,000 sq.ft. = 25.60

7) Sheath Walls-
1,424 sq. ft. x 41 M.H./1,000 sq.ft. = 58.38

8) Finish-
a) Int.Trim: 256 ft.x 0.03 M.H./ft.= 7.68
b) Ext.trim: 326 ft. x 0.04 M.H./ft.= 13.04
c) Doors: 9 drs. x 2 M.H./dr.= 18.00
d) Cabinets: 24 ft.x 0.2 M.H./ft. = 4.80
e) Clean-up: = 8.00

TOTAL = 399.64

399.64 man hours x $15.00/hr.	=	$5,994.60
+ 9% workman's comp.	=	$ 539.51
TOTAL CARPENTRY LABOR	=	$6,534.11

MATERIALS COST (NON-LUMBER)

for 960 sq. ft. house as per plans

1. PLUMBING (waste, drain, & supply = $300; fixtures = $500	= $	800.00
2. ELECTRICAL & FIXTURES (960 sq.ft. x $.50/sq. ft.) + $200 fixtures	= $	680.00
3. HEATING (forced air furnace & ducts)	= $	960.00
4. SHEET ROCK (3488 sq.ft. x $0.16/sq.ft.)	= $	558.08
5. ROOFING (12.8 squares (10'x10') x $45/sq.)	= $	576.00
6. CABINETS (uppers: 11 ft. x $31/ft.) (lowers: 12 ft. x $32/ft.)	= $	1,045.00
7. PAINT (exterior: 1424 sq.ft. x $0.06/sq.ft.) (interior: 3488 sq.ft. x $0.04/sq.ft.)	= $	224.78
8. COUNTERTOPS (22 l in.ft. x $8/ft.) + $7/sink + $12/end	= $	219.00
9. TILE (over tub (50 sq.ft. x $.50/sq.ft.)	= $	250.00
10. WINDOWS (9 x $70 each)	= $	630.00
TOTAL	= $	5,942.86

SUBCONTRACTORS AND SUPPLIERS REQUIRED TO COMPLETE ESTIMATE AND COST EVALUATION

500	Grading and Excavating
2,800	Plumbing
2,155	Heating and Sheet Metal
2,200	Electrical and Electrical Fixtures
1,045	Cabinets (installation is included in estimate)
2,220	Sheet Rock Hanging and Finishing
1,300	Roofing
675	Glass Windows and Doors
1,550	Finished Floors
150	Shower Door
—	Mirrors and Glazing
550	Tile
2,251	Painting
200	Weatherstripping
219	Countertop
—	Finish Hardware - Door Latches, Towelbars, etc.
—	Appliances
300	Concrete Flatwork - Exterior
18,096.00	TOTAL

Appendix C

This chart compares two local estimating guides (Bank of America and Hudson Home Estimator) with two real projects. This will give you a good idea of how much reality can veer from projected estimates.

COST ANALYSIS OF DIFFERENT PROJECTS

CONSTRUCTION ACTIVITY	B of A COST STUDY		HUDSON HOME ESTIMATOR		960 S.F. HOME		ALBANY ROOM ADDITION	
	% of Total	$ Cost per S.F.	% of Total	$ Cost per S.F.	% of Total	$ Cost per S.F.	% of Total	$ Cost per S.F
Permits	1.70	0.58	1.21	0.53	1.53	0.52	1.04	0.66
Site Preparation	1.10	0.40	0.80	0.35	1.23	0.42	0.07	0.44
Concrete Prep.	2.30	0.82	2.04	0.90	0.92	0.31	5.72	3.62
Foundation	3.60	1.28	6.16	2.72	6.03	2.14	7.18	4.54
Rough & Finish hardware	0.60	0.24	1.50	0.66	2.67	0.91	1.28	0.81
Lumber-Rough & finish	12.00	4.19	19.50	8.60	10.90	3.70	11.39	7.21
Carpentry Labor-TOTAL	9.50	3.30	8.36	3.69	11.22	3.81	20.77	13.14
Cabinets	3.30	1.16	2.60	1.14	4.30	1.46	1.74	1.10
Insulation	2.20	0.77	1.85	0.82	1.50	0.51	2.23	1.41
Roofing	6.70	2.36	7.75	3.42	3.99	1.35	2.49	1.57
Doors	2.10	0.72	4.86	2.14	1.90	0.64	0.51	0.32
Windows	1.30	0.45			2.07	0.70	2.67	1.69
Sheet rock: material, hanging & taping	4.40	1.54	6.25	2.76	6.75	2.29	6.74	4.26
Stucco/Siding	4.90	1.72	3.98	1.76	5.18	1.90	3.25	2.06
Flooring	4.30	1.50	5.29	2.33	4.45	1.51	3.88	2.46
Painting	3.60	1.27	3.44	1.52	6.44	2.19	4.36	2.76
Tile	0.40	0.13	---	---	1.69	0.57	4.00	2.53
Shower door, mirror & bath access	0.60	0.22	1.52	0.67	0.46	0.16	1.28	0.81
Plumbing	7.50	2.64	8.38	3.70	8.29	2.81	7.13	4.51
Heating & sheet metal	3.80	1.34	5.02	2.22	6.14	2.08	2.40	1.51
Electrical	3.90	1.38	7.08	3.12	6.75	2.29	5.26	3.33
Insurance	2.00	0.69	2.42	1.07	1.66	0.56	1.34	1.34
Plans & Specs.	0.50	0.16			2.00	0.68	1.86	1.18
TOTAL	82.30	28.86	100.00	44.12	98.07	33.51	100.00	63.29

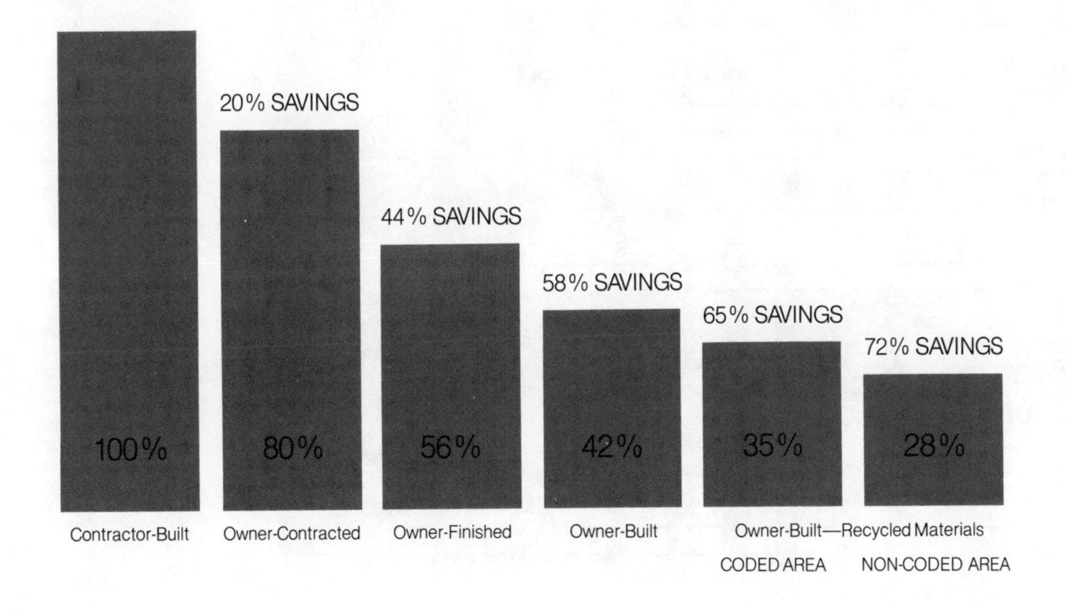

Appendix D

This chart is from Ken Kern's book *The Work Book: The Politics of Building Your Own Home*, and gives time estimates for each procedure. Again, like cost estimates, they can vary greatly.

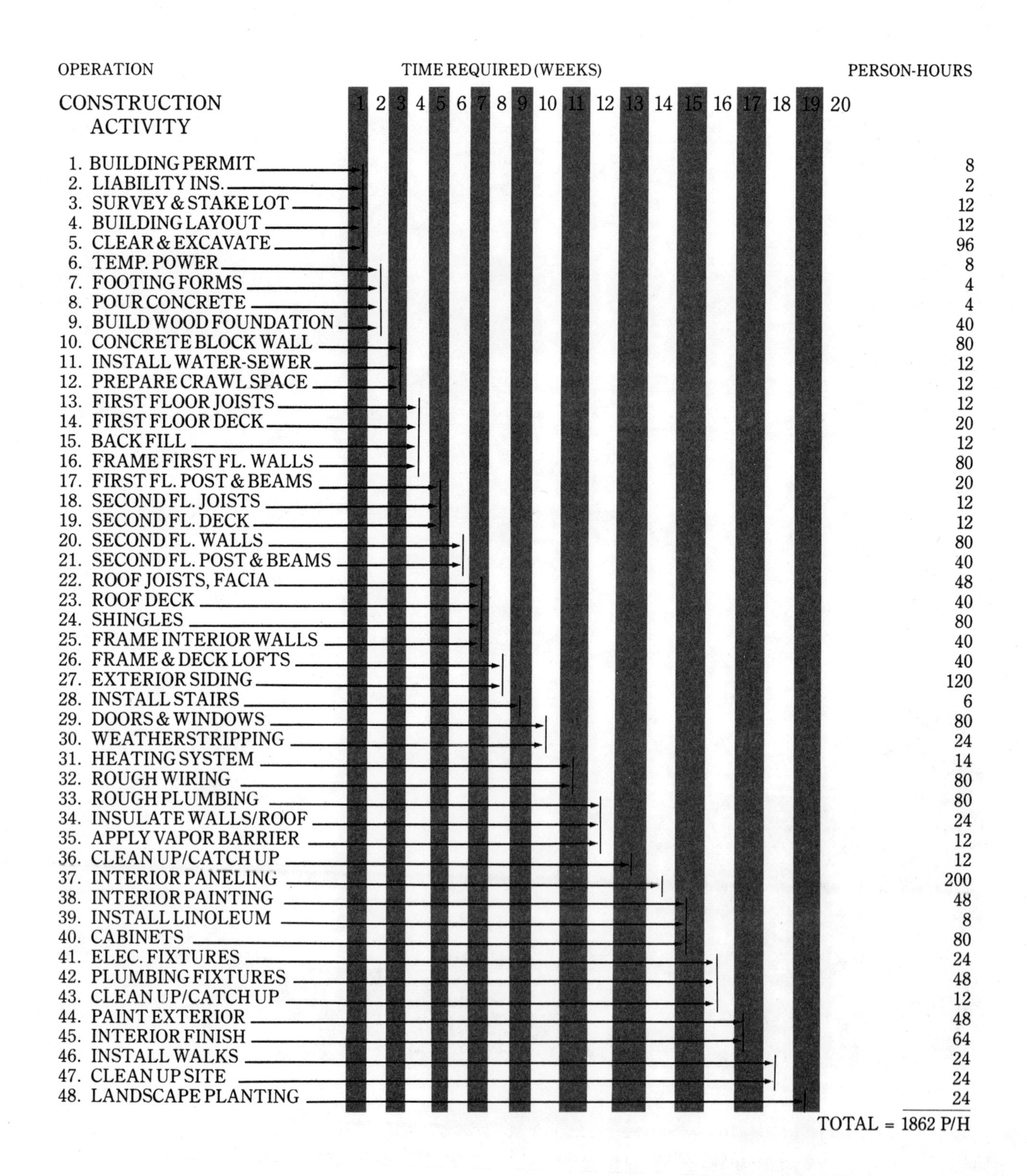

This page can be duplicated and used as a daily list once you start to build. It will help you organize each day as to the things you need to do or items you need to buy or rent.

Want List

Name ____________________ Day/Date ____________ Lot Number ______

ITEM	NEEDED ON JOBSITE Day/Date/AM-PM

Rental Tools*

Purchase Tools*

Purchase Materials*

Catalogs, Data Sheets, etc.

Design Decisions (Name of decision-maker—brief outline/explanation/question)

Inspections

Other

*List all important details: make, model, size, color, capacity, finish, type, grade, etc.
For specialty items list suggested vendors. Use reverse side for additional space.

The Owner-Builder Center

THE OWNER-BUILDER CENTER is a non-profit, self-funded educational organization located in Berkeley, California. We are dedicated to providing high quality information to the public on housebuilding, remodeling, energy efficiency, and solar energy. Founded in 1978, we have taught over 8,000 people in our classes and seminars and assisted many others in our consulting and design services. We have a highly qualified and dedicated teaching staff of thirty part-time teachers and five full-time employees who do the work of ten and still smile. Our classes are both local and national in scope.

Local Classes and Services

Locally, we teach classes in ten San Francisco Bay Area cities in the following: housebuilding, remodeling, design, plan drawing, cabinet making and solar hot water system installation. We also maintain a bookstore, reference library, and consultation service to assist people with individual needs. We also offer a series of seminars four times each year and sponsor local hands-on workshops.

Summer Residence Program

Of interest to people outside our immediate area, we offer summer residence housebuilding programs in Grass Valley, California in the foothills of the Sierra Nevada Mountains. These two- and three-week courses run between May and September, and include both hands-on and classroom instruction. During each summer we build several structures, usually including at least one passive solar home. The structures are begun at various times during the summer to allow for a great variety of jobsite work during each semester. Evenings are free for social time, a guest speaker or additional classes. On weekends are raft trips, swimming, hiking or exploring the Mother Lode country. Students are housed and fed at a local Quaker high school with a beautiful three hundred acre campus. Both cabins and campsites are available.

Consulting Service

The Owner Builder Center operates a full-service consulting business specializing in the needs of owner-builders and passive solar and energy-effi-

Front Row: Bob Beckstrom, Blair Abee, John Hughes, Robert Roskind, R.C. *Second Row:* Dave Linebarger, Brian Spencer, Mary Schmidt, Dan Zelinger, Constance McKnight, Mike Hurley, Tom Hise, Dan Fuller. *Third Row:* Bob Clark, John Seltzer, Mike Iverson. *Fourth Row:* Rich Zarlow, Bob Brown, Doug Ballon. *Fifth Row:* Jeff Reed, Joseph Costarella, Rick Kattenberg, Chuck Rumwell. *Sixth Row:* John Reed, Demetrios Nichols, Curt Burbick, Dave Roberts, Mel Berry.

cient design. The service operates on an hourly fee basis and consultation can be done by mail or by phone for those out of the area. Plans, photographs of the site, topographic maps, etc., can be sent to us with the questions you would like to be advised on and we will send you back either written replies or cassette tapes with our recommendations. You can, for instance, send us a copy of your plans and we can assist you with the estimate, do a materials take off, evaluate the plan for solar and energy efficiency, evaluate its complexity for a novice builder, or just do a general critique from a design perspective. We can do a feasibility study on the project as well, to assist you in determining whether the money, time and energy you have allocated to the project are sufficient. We can also assist you in designing your own house, and our draftspeople can draw up your plans. This is somewhat more difficult to do by phone or mail and it is advised that you visit in our Berkeley office if at all possible.

Training for New Centers

The Owner Builder Center is interested in assisting other individuals and groups in opening centers similar to ours or teaching housebuilding or remodeling classes in their local areas. At present we have trained eleven groups around the country (see list below). Our training programs last seven to ten days and take place in our Berkeley office. The program includes training in advertising, graphics, renting classrooms, teaching skills, consulting, bookkeeping, media and public relations, enrollment procedure and hiring and training teachers. Also included is a set of 2,500 slides on housebuilding, including the following subjects:

passive solar, layout, foundations, sill plates, knee walls, floor framing, subfloors, wall framing, rafters, sheathing, siding, roofing, plumbing, electricity, windows and soffits. Aside from the groups we have trained, we also work closely with two other owner-builder schools, *Heartwood* in Washington, Massachusetts and *Cornerstones* in Brunswick, Maine.

Organizations trained by
The Owner Builder Center

Bomar Owner Builder Center
704 Gimghoul Road
Chapel Hill, North Carolina 27514
(919) 929-3243

Building Resources
Community Housing Investment Fund
121 Tremont
Hartford, Connecticut 06105
(203) 233-5165

Chico Housing Improvement Program
539 Fume Street
Chico, California 95926
(916) 342-0012

Colorado Owner Builder Center
P O Box 12061
Boulder, Colorado 80302
(303) 449-6126

Denver Owner Builder Center
5835 West 6th Avenue, Unit 4D
Lakewood, Colorado 80214
(303) 232-8571

Durango Owner Builder Center
P O Box 3447
Durango, Colorado 81301
(303) 247-2417

Fine Homes Unlimited
512 Mystic Way
Laguna Beach, California 92651
(714) 494-6411

Hanford Adult School
120 East Grangeville Road
Hanford, California 93231
(209) 582-4401

Matsu Community College
University of Alaska
P O Box 899
Palmer, Alaska 99645
(907) 745-4255

Michigan Owner Builder Center
1505 East Eleven Mile Road
Royal Oak, Michigan 48067
(313) 545-7033

Minnesota Owner Builder Center
2608 Bloomington South
Minneapolis, Minnesota 55407
(612) 721-8887

Northwest Owner Builder Center
1139 34th Avenue
Seattle, Washington 98122
(206) 324-9559

Owner Builder Center
of Fairfield and Westchester
335 Post Road West
Westport, Connecticut 06880
(203) 226-7095

Owner Builder Center
at Miami-Dade Community College
Environmental Demonstration Center
11011 South West 104th Street
Miami, Florida 33176
(305) 596-1017

Owner Builder Center of New York
160 West 34th Street
New York, New York 10001
(212) 736-4909

Owner Builder Center of Sacramento
P O Box 739
Fair Oaks, California 95629
(916) 961-2453

Owner Builder Center of Southern California
361 East Magnolia
Burbank, California 91502
(213) 841-1942

Pacific Owner Builder Center
4562 Aukai Avenue
Honolulu, Hawaii 96816
(808) 523-8056

Tasmania Owner Builder Center
4 Dallas
Taroona, Tasmania
AUSTRALIA

Yestermorrow
P O Box 344
Warren, Vermont 05674
(802) 495-3437

If you would like to be on our mailing list for summer residence programs or would like more information about our consulting service, or starting classes or a center like ours in your area, please fill out the form below and mail it to our office (our mailing list is confidential and used only by the Center):

The Owner-Builder Center
1516 Fifth Street
Berkeley, California 94710
(415) 848-5951

Name ______________________________

Address ____________________________

______________________________ Zip ________

Store where book was purchased:

____

City and State: ______________________

☐ Please put me on your mailing list for summer residence programs

☐ Please send me information about your consulting service

☐ Please contact me about starting classes or a center

☐ Please put my friend on your mailing list:

Name ______________________________

Address ____________________________

______________________________ Zip ________

____

To order additional copies of this book, send $7.95 per copy plus $1.00 for handling (and 50¢ for each additional copy) to **The Owner Builder Center**, address above (California residents add 6%, Bay Area residents add 6½% sales tax). Inquire about quantity discounts.

Building Your Own House

This series is a complete owner-builder's manual for building a code-endorced passive solar, energy-efficient house. Each chapter deals with the tools and materials you will need, how long each process will take, what it will be like, the code requirements at each stage, most common mistakes, tricks of the trade, alternatives, and a complete step-by-step description of how each process is done, so that it can be accomplished by a novice builder. Each volume is profusely illustrated and easy to understand.

Volume I includes: laying out the foundation, foundations, sill plates, knee walls, floor framing, subfloors, wall framing, roof framing, sheathing, siding, roofing, and window and door installation.

Volume II includes: plumbing, electricity, heating and air conditioning, insulation.

Volume III includes: interior walls, ceilings, flooring, finishing plumbing and electricity, trim, and decks.

The Owner Builder Center is interested in assisting other individuals and groups in opening centers similar to ours or teaching housebuilding or remodeling classes in their local areas. At present we have trained twenty groups around the country (see list on the next page). Our training programs last seven to ten days and take place in our Berkeley office. The program includes training in advertising, graphics, renting classrooms, teaching skills, consulting, bookkeeping, media and public relations, enrollment procedure and hiring and training teachers. Also included is a full set of 3,500 slides on housebuilding, and 3,500 slides on remodeling, with which these classes are taught. For a free subscription to the *Owner Builder Newsletter,* including program dates and times, or new center locations, call (800) 547-5995. Other inquiries should be made to (415) 547-5995.

Reader Questionnaire

IN ORDER TO assist us in later editions of this book, your feedback and suggestions would be greatly appreciated. If you have a moment to fill out the following questionnaire, whether you have begun building or not, it will be a value to us and other owner-builders down the way.

Name and Address (optional) ____________________

City ____________________ State ____________________

Reason for purchasing book:

☐ Plan to build
☐ General interest
☐ Other ____________________

Any suggestions, additions or corrections, either in general or for any detail?

For you, what were the strengths of the book?

The weaknesses?

Did the workbook format work well for you? ☐ YES ☐ NO

Other chapters or areas you would like to see included:

Would you like to be on our mailing list for future books? ☐ YES ☐ NO

Name ____________________

Address ____________________

____________________ Zip ________